Progress in Theoretical Computer Science

Hava T. Siegelmann

Neural Networks
and Analog Computation
Beyond the Turing Limit

Birkhäuser
Boston • Basel • Berlin

Hava T. Siegelmann
Department of Information Systems Engineering
Faculty of Industrial Engineering and Management
Technion
Haifa 3200, Israel

Cover illustration by Pepi Marzel.

Library of Congress Cataloging-in-Publication Data
Siegelmann, Hava T.
   Neural networks and analog computation : beyond the Turing limit / Hava T.
Siegelmann.
      p.   cm. — (Progress in theoretical computer science)
   Includes bibliographical references and index.
   ISBN 0-8176-3949-7 (hardcover : alk. paper)
   1. Neural networks (Computer science)   2. Computational
complexity. I. Title. II. Series.
QA76.87.S565 1999
006.3'2—dc21   98-29446
                                         CIP

AMS Subject Classifications: 68Q05, 68Q15, 68T

Printed on acid-free paper.
©1999 Birkhäuser Boston   *Birkhäuser*

ISBN 0-8176-3949-7
ISBN 3-7643-3949-7

Formatted from the author's LaTeX files.
Printed and bound by Braun-Brumfield, Inc., Ann Arbor, MI.
Printed in the United States of America.

9 8 7 6 5 4 3 2 1

Contents

Introduction

Humanity's most basic intellectual quest to decipher nature and master it has led to numerous efforts to build machines that simulate the world or communicate with it [Bus70, Tur36, MP43, Sha48, vN56, Sha41, Rub89, NK91, Nyc92]. The computational power and dynamic behavior of such machines is a central question for mathematicians, computer scientists, and occasionally, physicists.

Our interest is in computers called *artificial neural networks*. In their most general framework, neural networks consist of assemblies of simple processors, or "neurons," each of which computes a scalar activation function of its input. This activation function is nonlinear, and is typically a monotonic function with bounded range, much like neural responses to input stimuli. The scalar value produced by a neuron affects other neurons, which then calculate a new scalar value of their own. This describes the dynamical behavior of parallel updates. Some of the signals originate from outside the network and act as inputs to the system, while other signals are communicated back to the environment and are thus used to encode the end result of the computation.

Unlike the von Neumann computer model, the structure of a neural network cannot be separated into a memory region and a processing unit. In the neural model, memory and processing are strongly coupled. Each neuron is part of the processing unit, and the memory is implicitly encoded in the mutual influence between any pair of neurons. The influence may be excitatory; its strength, then, will be measured by a positive real weight of the link. Inhibitory influence will be represented by a negative real number weight.

The status of the weights, whether perceived as unknown parameters or fixed constants, prompts two different views of the neural model. When the weights are considered unknown parameters, the network is a semi-parametric statistical model, able to approximate input-output mappings by means of parameter estimation (also called *learning* or *adaptation*). When the weights are considered constant (after or without a process of adaptation) the networks can perform exact computations rather than mere approximations. Due to the rich and numerous adaptation processes available, neural networks are of great use as nonlinear adaptive controllers and function approximators in industry.

The first computational model of a biological neuron was suggested by

McCulloch and Pitts. In their paper, entitled "A Calculus of Ideas Immanent to Nervous Activity" [MP43], they argued that because of the all-or-nothing characteristic of action potentials (as seen by Hodgkin and Huxley), the nervous system could be thought of as a finite interconnection of logical devices. Nervous activity could thus be interpreted as forming relations between symbolic propositions, which then proceed to build propositions of yet greater complexity. The McCulloch-Pitts neuron is still considered a classical digital abstraction of a simplified brain cell.

This work by McCulloch and Pitts was first presented in 1943, in the second annual meeting of the Josiah Macy Jr. scientific conference. Attended by biologists, mathematicians, engineers, and social scientists, the goal of the conference was to identify applicable parallels between brain and computer. The theory of McCulloch and Pitts was controversially received: many stressed that while digital behavior is a recognized component of cerebral functioning, any theory that did not take into account continuous variables (i.e., the underlying chemical and physical phenomena) could not be considered an accurate model. From the debates and discussions, two different approaches emerged. The first, lead by Gerard and Bateson, emphasized the significance of continuous-valued computation; the second, headed by McCulloch, Pitts, and von Neumann, stressed the advantages in the simplicity of digital computation [Nyc92].

Following the development of von Neumann's universal model of computation based on the principle of the McCulloch-Pitts neuron, the digital approach prevailed in the field of cybernetic research. The digital model laid the groundwork both for the 20th century's computer paradigm and for an enhanced understanding of computational processes in the brain.

With a backward glance to the 1940's, this book returns to the less popular principle of analog computation, from which we develop the foundations of continuous valued neural networks. When considering analog computers, the model of the neural network is not unnatural: while in 1943 neural networks were envisioned as digital, the adaptive neural models used for industrial purposes in the 1980's were, in fact, analog.

The theoretical foundations presented herein concern a particular kind of neural model: an *analog recurrent* neural network (ARNN). It consists of a finite number of neurons arranged in a general architecture. The network is analog in that it updates continuously and utilizes a continuous configuration space (as opposed to the discrete models of digital computers). In our analog model, each neuron is characterized by real constant weights and can hold a real value updated by means of a continuous function of its combined inputs. The adjective "recurrent" emphasizes that the interconnection is general, rather than layered or symmetrical. Such an architecture allows for internal state representation and describes its evolution over time.

We interpret the dynamical behavior of a network as a process of computation, and study the effects of various constraints and different parameters on its computational power. We demonstrate that the network is indeed a parametric model of computation: altering its constitutive parameters allows the model to coincide with previously considered models that are computationally and conceptually different. We prove that our network is inherently richer than the standard computational (Turing) model: it encompasses and transcends digital computation, while its power remains bounded and sensitive to resource constraints. Furthermore, it encompasses many analog computational models. This leads us to propose, analogous to the Church-Turing thesis of digital computation, a thesis of analog computation which states: No possible abstract analog device can have more computational capabilities than ARNN.

This thesis is a point of departure for the development of alternative (super-Turing) computational theories, while it relates to the Turing theory in a variety of ways. We first note that the assumption of continuity may be beneficial, even when the hardware is digital. A prime example is the linear programming problem, with its polyhedral structure of solutions [Kar91, PS82]. A search in the space of vertices by the simplex algorithm has exponential worst case behavior. On the other hand, interior point algorithms, such as Karmarkar's algorithm, approach the solution from the inside of the continuous polytope and require polynomial time only. This exemplifies the speedup gain from considering a continuous phase space.

However, the case is different if we assume a non-digital, nature based hardware which handles real values at once, disregarding their bit representation. Chaotic systems, for example, are inherently analog and are sensitive to minute changes in their parameter values. Digital computers can only approximate chaotic systems up to a certain point in time, according to the precision provided for the initial description. Our theory supports the argument that algorithms that are based on real numbers and act in a continuous domain not only provide a speedup but also describe natural phenomena that cannot be mimicked by digital models.

Another way to expand on the Turing model is to consider our theory as describing an extension not from digital to analog, but rather from static to adaptive computational schemes. This view is grounded in the finding that although the neural model is based on real numbers, the actual number of bits affecting the output is a monotone function of the computation time. The interleaving of computation and weight adaptation in a single machine enables the model to outperform the static Turing model in terms of its computational power.

In the 1940's, Claude Shannon suggested the General Purpose Analog computer (GPAC) as a theoretical formalization of continuous time analog computers [Sha41]; these were very popular at the time (e.g., Bush's MEMEX machine [Bus70]; later prominent models include [Pou74, PR88, Rub81, Rub89, Rub93]). Although they have "analog" in their title, these models are very different from the one proposed here. Shannon processors are integrators, i.e., they calculate integrals, whereas our processors are continuous functions applied to continuous values.

Organization of the Book

It is our goal to make the research in the area of computational complexity of neural networks accessible to a wide range of readers, not only mathematicians and computer scientists, but also general scientists and engineers, who are interested in the computational foundations of neural networks or in analog computation. Each chapter begins with a brief introduction which explains the particular ideas therein; some chapters also end with notes that provide complementary details. Initially, the reader may choose to focus primarily on the introductions and skip over some of the mathematical proofs, in order to obtain an overview of the book.

The book is organized as follows. In Chapter 1 we introduce basic terminology regarding computational complexity, thus providing the framework necessary for the computational understanding of the analog recurrent neural networks (ARNN), as described in Chapter 2. We are first interested in determining the type of functions computable by ARNN, obtained by enforcing different constraints on the network's constants. We show how the computational power of a network depends on the type of numbers utilized as its weights: by altering the type of weights to integers (Chapter 2), rationals (Chapter 3) and reals (Chapter 4), we produce networks that are computationally equivalent to finite automata, Turing machines (the mathematical equivalent of digital computers), and nonuniform computational models, respectively.

In Chapter 5 we lay out a hierarchy of computational classes between the digital and the analog machines. This is done using a variant of the Kolmogorov complexity measure of Information Theory; that variant considers both the information encoded in the input and the time complexity required to decode it as separate resources that effect the computational complexity.

We then depart from constraining the weights of a network, and investigate how other aspects determine the network's power. In Chapter 6, we constrain the precision allowed in the neurons. The resulting networks correspond to computational (Turing) space classes. We then monitor the computational power by modifying the nonlinear activation function that transfers the combined input stimuli of a neuron to a scalar output. Up to this point the function referred to was piecewise linear. In Chapter 7, we consider the smooth activation function that is nowadays very popular in the framework of neural networks as a semi-parametric model; this is the tanh, also called the *sigmoid*. The resulting network is computationally universal. In Chapter 8 we examine a very large class of activation functions that yield the power of finite automata.

So far we have dealt with deterministic networks only. In Chapter 9 we consider a generalization of the von Neumann model of an interconnection of unreliable components. It is an analog recurrent interconnection of stochastic neurons, whose random behavior may be characterized by a fixed value, or it may be influenced by the history and the environment. We prove that when utilizing only rational weights, the stochastic networks are computationally stronger than the Turing model, but still strictly weaker than the ARNN. This new super-Turing model of computation has similarities to the model of quantum computers, and may be considered feasible.

In the last three chapters of the book we develop a view of analog computing which is inspired by our network; we then compare it with other models of analog computation. We begin in Chapter 10 by demonstrating an equivalence in computational power between our network and that of a large class of generalized analog processor networks. We further prove that the networks of this class have some degree of tolerance to noise and architectural imprecisions. In Chapter 11, we survey models of analog computation and suggest a list of features characterizing physically realizable models. In the last chapter we proceed to prove that our network model encompasses many other analog computational models. This is the basis for the analog computation thesis that considers the neural network as a standard in the realm of analog computation.

Acknowledgments

The theory presented in this book originated from joint work with Eduardo Sontag, to whom the book is dedicated with gratitude and admiration. Chapters 3, 4, 10, and part of 12, are based on the work done with Eduardo Sontag [SS91, SS95, SS94]. Chapters 5 and 6 are based on work with José Balcázar and Ricard Gavaldà [BGSS93, BGS97]. Chapter 7 is based on a paper with Joe Kilian [KS96]. I thank all of these colleagues for their permission to use the material here. Chapters 8, 9 and part of 12 are based on the papers [Sie96a], [Sie98] and [Sie95], respectively.

I thank my editor Lee Cornfield and my student Asa Ben-Hur for their help in transforming the rough draft of this book into a readable form. I also express appreciation to Felix Costa who generously provided suggestions and my assistant David Barnett for helping with the preparation of the manuscript.

My deepest gratitude to Hanna and Joseph Siegelmann whose love and support have been a constant source of my happiness and inspiration. Finally, I acknowledge my debt over the years to my late grandparents Abraham and Tova Siegelmann, Max and Rachel Lowenstein, and my late mentor Yona Benedikt.

Chapter 1

Computational Complexity

Although neural networks are based on continuous operations, we still analyze their computational power using the standard framework of computational complexity. In this chapter we provide the background material required for the search of the computational fundamentals of neural network and analog computational models. Our presentation starts with elementary definitions of computational theory, but gradually builds to advanced topics; each computational term introduced is immediately related to neural models.

The science of computing deals with characterizing problems and functions. A function, in the mathematical sense, is a mapping between the elements of one set called the *domain*, and the elements of another called the *range*; consequently, a function can be thought of as a set of domain-range ordered pairs. If the function is defined for each domain element, then it is said to be *total* (also called a *map*), otherwise it is *partial*. Unless otherwise specified, in this book we consider partial functions only. Two partial functions with the same domain and range are said to be equal if, for every domain element, either the two functions are undefined or they are of equal value.

A computable function is a unifying rule between "computable objects" (i.e., objects that can be explicitly presented by finite means, like natural, integer or rational numbers) that specify how to get the second element from the first. A noncomputable function is an infinite set of ordered pairs (of computable objects) for which no "reasonable" rule can be provided. The classical theory of computation is focused on both domains and ranges that are discrete (rather than continuous). The finiteness (discrete domain) of the input and output is a crucial requirement in the theory of computing, as it ensures that the power of a model is based purely on its internal structure, rather than on the precision of the environment. When the range is binary, the functions are called *characteristic functions* or *indicator functions* of predicates. In this case, the collection of all domain elements that are mapped to "1" constitutes a *language*. For all computational purposes, functions and languages are equivalent, as will be explained in Remark 1.8.1 below; we thus use one or the other interchangeably.

The field of computational complexity is primarily dedicated to partitioning sets of functions into a partially ordered hierarchy of computational classes with an increasing degree of difficulty. Historically, in automata theory, functions were characterized by the type of automata that compute them, or in other words, each computational class of functions was associated with such an automaton. Automata are theoretical machines, also called *computational models* and *computational machines*. We say that one type of automata is *stronger* than another if the first can *compute* (or *generate*) a set of functions or *decide/recognize* (or *accept*) a set of languages, while the latter can compute only a strict subset of these functions or languages. The most popular automaton is the Turing machine; it is neither the weakest nor the strongest, but it is the mathematical equivalent of a digital computer having unbounded resources [HU79, BDG90].

The modern theory of computational complexity does not deal only with the ultimate power of a machine, but also with its expressive power under constraints on resources, such as time and space. Resource constraints are defined as follows. To each domain element ω we associate a measure $|\omega|$ called its *size* or its *length*. We then define a partial function on natural numbers $T : \mathbb{N} \to \mathbb{N}$. $T(|\omega|)$ is defined only when the computation of all domain elements of this size halt. We say that a machine $\mathcal{M}$ *computes in time* T, if for all inputs ω for which $T(|\omega|)$ is defined, $\mathcal{M}$ halts after performing not more than $T(|\omega|)$ steps of computation. Time constraints give rise to *time complexity classes* of functions or languages, i.e. the class A-Time $(T(n))$ consists of all the functions/languages that are computed/decided by some automaton in A in time $T(n)$. Similarly, space constraints (i.e. the number of cells scanned during the computation) give rise to *space complexity classes*. We will concentrate mainly on time complexity classes, and refer to them only as complexity classes, with the notation $A(T(n))$.

The rest of this chapter is organized as follows. We start with a brief introduction of neural networks; we then present an overview of some types of automata and their complexity classes and discuss their relation with neural computation. Our presentation begins with the weakest model and proceeds to more powerful ones, concluding with the advice Turing machine, which is the most powerful and pertains most directly to our work.

1.1   Neural Networks

The basic paradigm in neural networks is that of an interconnection of simple processors, or "neurons." The directed connection between any two neurons is characterized by a positive or negative real number, called a *weight*, which

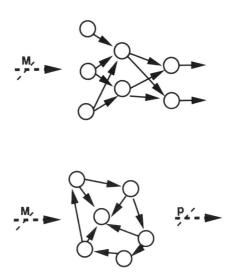

Figure 1.1: Feedforward (above) and Recurrent Networks (below)

describes the relative influence. Each neuron computes a scalar function of its weighted inputs. This scalar function, also called *activation function* or *response*, imitates the monotonic response of biological neurons to the net sum of excitatory and inhibitory inputs affecting them. The output produced by each neuron is, in turn, broadcast to other processors in the network. In this book we will use the terms "neural network" and "processor network" interchangeably.

In general, one may classify neural networks according to their architecture, into *feedforward* and *feedback* (or *recurrent*) networks. The former are arranged in multiple layers, in such a manner that the input of each layer is provided by the output of the preceding one. The interconnection graph is acyclic, and the response time to an external input cannot be greater than the number of layers, independent of the length of the input. Feedforward networks are useful for the representation, interpolation, and approximation of functions and stationary time series (see, e.g., [Bar92, BH89, Cyb89, Fra89, Hor91, HSW90, MSS91, Sus92, SW90, Son93]), but because their computation ends in a fixed number of steps, one such network cannot perform general computations for inputs of varying lengths. Feedback networks, in contrast, allow loops in their graphs, and thus memory of past events is possible; this property facilitates a more compact and general representation of time series (see, e.g., [Son93]). These networks can be considered a rich computational model as we will see in the following chapters.

1.2   Automata: A General Introduction

The search for mathematical models that reflect various physical control systems began in the field of automata theory [Min67]. The components of the actual system may take many physical forms, such as gears in mechanical devices, relays in electromechanical ones, integrated circuits in modern digital computers, or neurons. The behavior of such systems depends on the underlying physical principles. The description of a system as an automaton requires the identification of a set of *states* that characterize the status of the device at any moment in time, and the specification of transition rules that determine the next state based on the current state and inputs from the environment. Rules for producing output signals may be incorporated into the model as well.

Although automata were formalized prior to the advent of digital computers, it is useful to think of automata as describing computers, in order to explain their basic principles. In this view, the state of an automaton, at a given time t, corresponds to the specification of the complete contents of all RAM memory locations, and also of all other variables that can affect the operation of the computer, such as registers and instruction decoders. We use the symbol $x(t)$ to indicate the state of all variables at time t. At each instant (time step) the state is updated, leading to $x(t+1)$. This update depends on the previous state, instructed by the program being executed, as well as on external inputs, such as keyboard strokes and pointing-device clicks. We use the notation $i(t)$ to summarize the contents of the inputs. (It is mathematically convenient to consider "no input" as a particular type of input.) Thus one postulates an update equation of the type

$$x(t+1) = f(x(t), i(t)) \qquad (1.1)$$

for some mapping f, or in short-hand form, $x^+ = f(x, i)$, where the superscript "+" indicates a time-shift.

Also, at each instant, certain outputs are produced: update of the video display, characters sent to a printer, and so forth; $y(t)$ symbolizes the total output at time t. (Again, it is convenient to think of "no output" as a particular type of output.) The mapping h calculates the output at time t given the internal state at that instant

$$y(t) = h(x(t)) . \qquad (1.2)$$

Abstractly, an automaton is defined by the above data. Thus, as a mathematical object, an *automaton* is simply the quintuple

$$M = (Q, I, Y, f, h)$$

consisting of sets Q, I, and Y (called respectively the state, input, and output spaces), as well as two functions

$$f : Q \times I \to Q, \quad h : Q \to Y$$

(called the next-state and the output maps, respectively). The sets I and Y are typically finite.

When defining the input/output map (I/O map for short) produced by an automaton, the input set is

$$I = \Sigma \cup \{\$\}$$

where Σ is the set of possible input letters, and $\$$ is a special letter designating the end of a string, or equivalently, the empty letter (not to be confused with the "no-input" of the RAM example that is a letter in Σ). Attention is constrained to inputs of the form $i = \omega\$^\infty$, where "$\omega$" is a finite sequence over Σ ($\omega \in \Sigma^*$) and "$\$^\infty$" is the infinite sequence of $\$$'s. Given such a finite input string and an initial state, a well-defined output string is obtained by recursively solving the update Equations (1.1) and reading-out the corresponding outputs. There are many possible conventions regarding the interpretation of the input/output behavior (map) of an automaton. The response may be defined as the last output symbol produced when the input sequence ends. Alternatively, the computation may end when a special ("accepting" or "final") state has been reached, and in this case the output can be defined as either the output letter generated at the moment of arrival, or as the sequence of output letters accumulated during the computation. In all cases, the output string is either finite or not defined.

1.2.1 Input Sets in Computability Theory

The input of digital computational models, and of the model described in this book, is a stream of symbols (traditionally called *words*, *strings*, or *sequences*) belonging to a finite non-empty set Σ, commonly called an *alphabet*. In what follows we will consider the *binary* alphabet, namely the set $\Sigma = \{0, 1\}$. Using more than two symbols yields at most a linear speedup, while using only one symbol may cause an exponential slow-down. Yet, in some particular cases, we will still concentrate on the single symbol (*unary*) alphabet $\Sigma = \{0\}$. In these cases the sets of domain elements that are mapped to "1" are not called languages but *tally sets* instead.

The following paragraph introduces the notation for some commonly used sets of words over the alphabet Σ. Σ^* denotes the set of all finite words over Σ:

$$\Sigma^* = \bigcup_{n \in \mathbb{N}} \Sigma^n ,$$

where Σ^n is the Cartesian product $\Sigma \times \cdots \times \Sigma$ of length n. By $\Sigma^{\leq n}$ we denote the set

$$\bigcup_{k \leq n} \Sigma^k .$$

Σ^* includes the empty word ϵ; Σ^+ is the set $\Sigma^* \backslash \{\epsilon\}$. In contrast to Σ^* and Σ^+, the set Σ^∞ of infinite sequences (one-way infinite) and the combined set $\Sigma^\# = \Sigma^* \bigcup \Sigma^\infty$ are not considered input sets.

All of the sets described above can be made ordered: strings will be ordered first by length and then lexicographically within each length.

1.3   Finite Automata

A *finite automaton* (FA) has a finite state space. Such an automaton is commonly described, e.g. in [Min67], as follows: The machine starts from a designated initial state. At each step it reads an input symbol and changes its state in accordance with a specific transition rule. The input is accepted if, after reading the last symbol, the machine is in one of the designated accepting states; otherwise the input is rejected. Other formulations exist, see [HU79].

For the formal definition, we deviate from the notation used above for general automata and adhere to the particular notation common in the literature. A FA is defined as a quintuple ([HU79]) $\mathcal{M} = (Q, \Sigma, f, q_0, F)$ where Q is a finite set of *states*, Σ is a finite set of *input* symbols also called the *alphabet*, $q_0 \in Q$ is the initial state, $F \subseteq Q$ is the set of *accepting* states, and $f : Q \times \Sigma \to Q$ is the *transition* function. That is, for any state q and input symbol a, we interpret $f(q, a)$ as the "next-state". We assume that $f(q, a) = q$ for any accepting state $q \in F$.

The single-step transition function, f, can be extended to define a transition function, f^*, mapping a state and a string of symbols to the resulting state. The inductive definition is as follows: if ω is the empty string ϵ then $f^*(q, \omega)$ is q, else if ω is $\omega' a$, then $f^*(q, \omega)$ is $f(f^*(q, \omega'), a)$, for a symbol $a \in \Sigma$ and strings $\omega, \omega' \in \Sigma^*$.

A string ω is said to be *accepted* by a finite automaton $\mathcal{M}$ if $f^*(q_0, \omega) \in F$. The *language* $L(\mathcal{M}) \subseteq \Sigma^*$ accepted by $\mathcal{M}$ is the set of all accepted strings. A language is *regular* if it is accepted by some finite automaton. Note that the decision to accept or to reject a string is made immediately after it is read.

1.3.1   Neural Networks and Finite Automata

Since the work by McCulloch and Pitts [MP43], mathematical modeling of neurons in theoretical computer science has been based on a binary response (activation) function, typically on the function signal(x) which is equal to 0 if $x \leq 0$ and is equal to 1 otherwise. A similar common function is the *Heaviside* function (also called *threshold*) $\mathcal{H}(x)$ which outputs 1 also for the value $x = 0$. A finite network of such two-state elements cannot accomplish more complex computations than those performed by finite automata.

It might appear that in practice it is sufficient to restrict our studies to finite automata. After all, the world is finite and only a finite amount of memory is available in any realizable machine. However, even for digital computation, finiteness imposes theoretical constraints that are undesirable when one is interested in understanding ultimate computational capabilities.

As an illustration, assume that one wishes to design a program that will read a finite binary input string and detect whether the number of "1"'s in the sequence is greater than the number of "0"'s. There is no machine that can memorize the difference between the occurrences of "0"'s and "1"'s of any string, using only a fixed predetermined memory (see Note 1). On the other hand, one could certainly write a computer program in any modern programming language to perform this task. The program would count the extra "1"'s and execute correctly as long as enough external storage (e.g., in the form of disk drives) is potentially available. We conclude, then, that finite automata are useful when modeling restricted programs or physical devices (e.g., lexical analyzers in compilers or sequential circuits), but they are not sufficiently powerful when developing a general theory of computation, which is our interest here.

A possible way to avoid the limitations of finiteness while still employing a threshold response is the use of *infinite* networks, i.e., an infinite number of neurons, although perhaps with only a finite number active at any given instant. (For computations with (potentially) infinite networks the reader can consult [FG90, GF89, HS87, Hon88, Orp94, Wol91]). These models, which use an unbounded number of computational units, do not comply with the framework used in the field of computational complexity, in which the processing unit should be bounded, and infiniteness may appear only in resources such as time and space.

Another way to overcome the finiteness of networks is to consider other activation functions that allow for continuous rather than binary values in the neurons. In this way, the number of neurons (processing elements) can be kept finite while the amount of memory (precision) is theoretically unlimited. Such a model is computationally rich, and yet it is also sensitive to resource and parameter constraints; thus its computational properties are of interest. This is the type of model we choose to study in this book.

Such a neural network, characterized by a finite number of processing elements combined with unbounded memory, brings us to the next model of computation.

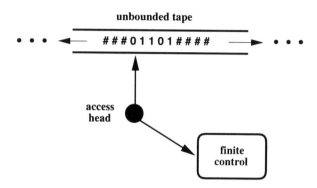

Figure 1.2: The Turing Machine Model

1.4   The Turing Machine

A mathematical model more general than that of finite automata, the *Turing machine*, allows for "external" storage, in addition to the information represented by the current "internal" state of the system. This model, introduced by the English mathematician Alan Turing in 1936, forms the basis for much of theoretical computer science (see, e.g., [BDG90]). In a Turing machine, a finite automaton is used as a "control" or main computing unit, but this unit has access to potentially infinite read/write storage space. The entire system, consisting of the control unit and the storage device, together with the rules that specify access to the storage, can be seen as a particular type of infinite automaton, albeit one with a very special structure.

Formally, a Turing machine consists of a finite control and a binary tape, infinite in both directions, as depicted in Figure 1.2. The tape is accessed by a read-write head. At the beginning of the computation, an input sequence is written in binary code on the tape, surrounded by infinite sequences of blanks on both sides (empty symbols, denoted by "#" in the figure). The head is located at the leftmost symbol of the input string.

At each step the machine reads the tape symbol ($a \in \{0, 1, \#\}$) under the head, checks the state of the control ($q \in \{1, 2, \ldots, |Q|\}$), and executes three operations:

1. It writes a new binary symbol into the current cell under the head ($b \in \{0, 1, \#\}$).

2. It moves the head one step to either the left or to the right ($m \in \{L, R\}$).

3. It changes the state of the control ($q' \in \{1, 2, \ldots, |Q|\}$).

The transitions of the machine are described by a function $g(a, q) = (b, m, q')$. When the control reaches a special state, called the "halting state," the machine stops. The output of the computation is defined by the binary sequence written on the tape extending from the read/write head to the first # symbol on the right side of the head. Thus, the I/O map, or the function, computed by a Turing machine is defined in terms of the binary sequences on its tape before and after the computation. If this function is computable and total then it is said to be *recursive*; otherwise, if it is computable and partial, it is said to be *recursively enumerable*. If the range is $\{0, 1, \}$, the function is recursively enumerable when its 1 instances are computable and it is recursive when both its 0 and 1 instances are computable.

Any Turing machine $\mathcal{M}$ can be encoded by a finite binary sequence $\tau(\mathcal{M})$ using techniques developed by Gödel and Turing. Let $\rho_{\tau(\mathcal{M})} : \{0, 1\}^* \rightarrow \{0, 1\}^*$ be the function computed by a Turing machine $\mathcal{M}$. The function $\rho_U : \{0, 1\}^* \times \{0, 1\}^* \rightarrow \{0, 1\}^*$ is said to be *universal* if $\rho_U(\tau(\mathcal{M}), \omega) = \rho_{\tau(\mathcal{M})}(\omega)$. Turing showed that the function ρ_U is computable by a Turing machine [BDG90]. The existence of a machine that receives the encoding $\tau(\mathcal{M})$ together with another input string ω and simulates the behavior of $\mathcal{M}$ step by step on the input ω is like an existence theorem for a *general purpose digital computer*. A Turing machine that computes ρ_U is called a *universal* Turing machine.

Consider the set of Turing machine configurations. Since the tape is unbounded, the set of configurations is infinite and the Turing machine can be thought of as a particular type of automaton with an infinite state space. Yet, like finite automata, Turing machines also are limited in their computational power. While the number of functions with binary domains and ranges is uncountable, the class of functions that are computable by Turing machines is countable and can be enumerated. An example of a function known *not* to be Turing computable is the *halting* function. This is the decision problem defined as follows: given two words $\tau(M)$ and ω, decide whether the machine $\mathcal{M}$ would halt when starting with the input ω; or in other words, whether it would ever reach its halting state. In Note 2 we outline the proof for why this function is not Turing computable.

These limitations have been formalized into a basic principle in computing theory, known as the Church-Turing thesis (C-T for short). It states that no possible abstract digital device can have more capabilities (except for relative speedups due to more complex instruction sets or parallel computation) than Turing machines. The Turing machine indeed fits the informal intuitive notion of an algorithm; also, the Turing polynomial computation corresponds with many formal models of efficient computation. Furthermore, an "intuitively computable" function that is beyond the Turing limit is yet to be found. In the eighties Deutsch introduced a *physical* interpretation of the mathematical

C-T that reads as follows: no realizable physical device can compute functions that are not computable by Turing machines [Deu85].

Speedup can be obtained when more than one tape is allowed in the definition of a Turing machine. In these machines, there is still only one control, but several heads, one for each tape, and the tape transitions occur in parallel. Other variants of the standard Turing machine have been introduced in the literature. Among the most popular are machines that restrict the tape(s) to be infinite in one direction only, and stack machines, which will be presented in Chapter 3.

Time and space constraints are emphasized in the Turing model, and computability under these constraints has been extensively studied. Here we concentrate mainly on time complexity. There are various computer architectures and computer languages, and hence different ways of measuring time steps. However, when comparing the computation time necessary for any two digital computers to execute any program, the ratio is bounded by a polynomial. For this reason, an efficient computation is defined as one that requires polynomial time on a Turing machine.

The complexity class P is the class of functions $f : \Sigma^* \to \Sigma^*$ (and in particular of all languages over Σ) computed by a Turing machine in polynomial time in the length of the input; that is, the computation is complete and the machine halts in time cn^k, for some constant $c > 0$ and some integer k. EXP is the class of binary functions that may take exponential computation time on Turing machines; it is considered an inefficient class. There is a similar classification for space constraints, e.g. the class PSPACE is the space analog of the class P.

The time classes just described are only two of the interesting time complexity classes. In what follows $O(f)$ designates the set of all functions g for which there exists a constant c_g such that $g(n) \leq c_g f(n)$ for all $n > n_0$ and for some n_0. Of particular interest are time classes that adhere to the property of being *closed under* $O(\cdot)$. We say that a class $\mathcal{F}$ of functions is closed under $O(\cdot)$ if for every f and g, if $g \in O(f)$ and $f \in \mathcal{F}$, then $g \in \mathcal{F}$.

1.4.1   Neural Networks and Turing Machines

Naturally, the Turing model captures much attention in neural computability, even when regarding networks with a finite number of neurons. Pollack showed that Turing machines can be simulated by a certain model, in which each neuron computes either a second-order polynomial or a thresholded affine function [Pol87]. Here we prove that by restricting attention to networks whose interconnection weights are all rational numbers and whose activation functions are homogeneous, continuous, and "simply-computable," we can obtain a model of computation that is polynomially related to the Turing model. In particular, we prove in Chapter 3 that any multi-tape Turing machine can be simulated in real-time by a network having rational weights. Furthermore, we present a network composed of fewer than one thousand neurons that computes all Turing computable functions preserving the time complexity. The converse simulation of the network by a Turing machine, with polynomial slow-down, is obvious.

We now introduce some additional models of computation.

1.5 Probabilistic Turing Machines

All types of Turing machines described above update deterministically. It is interesting to consider a model that is similar but allows the use of *random coins*: this is the basis of the operation of the probabilistic Turing machine and of the stochastic neural networks to be described in Chapter 9. In contrast to the deterministic machine, which acts on every input in a specified manner and responds in one possible way, the probabilistic machine may produce different responses for the same input.

Definition 1.5.1 ([BDG90], volume I): A *probabilistic Turing machine* is a machine that computes as follows:

1. Every step of the computation can have two outcomes, one chosen with probability p and the other with probability $1 - p$.

2. All computations on the same input require the same number of steps.

3. Every computation ends with *reject* or *accept*.

All possible computations of a probabilistic Turing machine can be described by a full binary tree (all leaves at the same depth) whose edges are directed from the root to the leaves. Each computation is a path from the root to a leaf, which represents the final decision. A coin, characterized by the parameter p, chooses one of the two children of a node to be the next one in the computational path. In the standard definition of probabilistic computation, p takes the value $\frac{1}{2}$.

The *error probability* of a probabilistic Turing machine $\mathcal{M}$ is the function $e_{\mathcal{M}}(\omega)$, defined by the ratio of computations on input ω resulting with the wrong answer to the total number of computations on ω (which is equal to $2^{T(|\omega|)}$ for computation time T). Probabilistic computational classes are defined relative to the error probability. PP is the class of languages accepted by polynomial time probabilistic Turing machines with $e_{\mathcal{M}} < \frac{1}{2}$. A weaker class defined by the same machine model is BPP, which stands for *bounded error probabilistic polynomial time*. BPP is the class of languages recognized by polynomial time probabilistic Turing machines whose error probability is bounded above by some positive constant $c < \frac{1}{2}$. The latter class is recursive, but it is unknown whether it is strictly stronger than P. There are other probabilistic classes such as R and ZPP, which are outside the scope of this book, see e.g., [BDG90].

1.5.1 Neural Networks and Probabilistic Machines

We introduce in Chapter 9 a particular model of stochastic neural networks. If we consider rational weights and $p = \frac{1}{2}$, we obtain the standard Turing probabilistic classes. Naturally, we do not restrict the value of p to $\frac{1}{2}$ but instead allow the neurons to behave stochastically according to a set of coins that are characterized by different values in $(0, 1)$.

If all p's are rationals, no change in the computational power is achieved. An interesting phenomenon occurs when p can take real values, keeping the weights rational. (The coins are still binary in values, only the probabilities are real.) Such networks compute functions that are beyond the Turing limit. This model can thus be considered a super-Turing model of computation, but it is still weaker than the deterministic neural network with real weights. Unlike the case of rational weights, when real weights are involved the coins add no extra computational power to the deterministic dynamics.

1.6   Nondeterministic Turing Machines

Nondeterministic operation is a popular way of extending the Turing machine. As in the probabilistic computation, any step of the nondeterministic machine makes one of two choices regarding what to write on the tape ("0" or "1"), how to move the head (left or right), and to which state of the finite control to proceed. An input sequence ω is said to be accepted by a nondeterministic Turing machine if there is *some* sequence of choices that will make the machine output "1" for the input ω. NP (*Nondeterministic Polynomial time*) is the class of functions that can be computed by some nondeterministic Turing machine in polynomial time. A well-known open research question is whether the classes P and NP coincide (see, e.g., [GJ79]); it is considered by some researchers to be the most significant mathematical problem of the twentieth century [Sma91].

An alternative description of the class NP is demonstrated by the following example. Let L be the language of all the binary representations of composite numbers. This language is hard to decide, but it becomes easy if in addition to an input n one is given two natural numbers n_1 and n_2, with the promise that n is composite if and only if the multiplication $n_1 n_2$ is equal to n. In this setup we call n_1 and n_2 the *certificate* of the "yes" answer to the question of whether

n is composite. The authentication of the certificate is all that is required in the NP framework. Here the authentication process is the multiplication $n_1 n_2$ and the comparison of the result with n. "No" instances have no authentic certificate.

This example can be formulated into a definition. A decision problem L is in NP if there exists a deterministic Turing machine that receives as input a pair $\langle \omega, \alpha(\omega) \rangle$, where $\alpha(\omega)$ is polynomially long in $|\omega|$, and accepts $\langle \omega, \alpha(\omega) \rangle$ in polynomial time if and only if $\alpha(\omega)$ is a certificate for ω. An example of a certificate is the series of choices in the nondeterministic model that lead the machine to output "1".

1.6.1 Nondeterministic Neural Networks

Nondeterministic computation in neural networks is briefly discussed in Chapters 3, 4, and 9. It is defined analogously to the definition of the class NP of the Turing machine.

We are interested both in classical/"weak" nondeterministic computation, where the certificate α is a string of binary digits, and in the "strong" nondeterministic computation suggested in [BSS89], in which the certificate is a series of real numbers. We assert in Chapter 9 that the "strong" and "weak" nondeterministic neural networks are computationally equivalent.

1.7 Oracle Turing Machines

Oracle Turing machines are more powerful than standard Turing machines. An oracle Turing machine has a specially designated oracle tape and three special states called QUERY, YES, and NO. When the machine enters the QUERY state, it switches in the next step to state YES or state NO, depending on whether the string currently written on the oracle tape is in an *oracle set* A, fixed for the computation. The resulting computation is relative to a set A, giving rise to *relativized complexity classes*. Note that the oracle does not have to be Turing computable, and that the answer to a membership query requires one unit of time, independent of the oracle. For an oracle set A, $\mathrm{P}(A)$ is the relativized class of languages that can be decided in polynomial time by oracle Turing machines querying the oracle A. If the language B is in $\mathrm{P}(A)$, we also say that B (Turing-)reduces in polynomial time to A. For a class of languages $\mathcal{A}$, $\mathrm{P}(\mathcal{A})$ is the union of all $\mathrm{P}(A)$ for $A \in \mathcal{A}$.

1.7.1 Neural Networks and Oracle Machines

In Chapter 5 we will consider oracle Turing machines that use tally sets as oracles. The complexity of the tally oracles translates into a hierarchy of relativized complexity classes. In neural networks, the amount of information carried by the weights is shown to play an analogous role.

1.8   Advice Turing Machines

The following model is the most relevant for this book, because we will show that our neural model, although uniform, computes the same complexity class as this nonuniform model.

Nonuniform complexity classes are defined by the model of advice Turing machines [KL80], which, in addition to their input, receive also another sequence that assists in the computation. For all possible inputs of the same length n, the machine receives the same advice sequence, but different advice is provided for input sequences of different lengths. When the different advice strings cannot be generated from a finite rule (e.g. a Turing machine), the resulting computational classes are called *nonuniform*. The nonuniformity of the advice translates into noncomputability of the corresponding class. Noncomputability does not arise with the probabilistic/nondeterministic machines, even though these models also rely on a sort of external "advice." Whereas for nondeterministic machines each input looks for one possible "yes" path, and for probabilistic machines each input requires most paths to be "yes", here the computation is deterministic. In the advice machines, the length of the advice is bounded as some function of the input, and can be used to quantify the level of noncomputability.

Let Σ be an alphabet and let \$ be a distinguished symbol not in Σ; $\Sigma_\$$ denotes $\Sigma \bigcup \{\$\}$. We use homomorphisms $h^* : \Sigma_\$^* \mapsto \Sigma^*$ to encode words. These homomorphisms are inductively defined in terms of an encoding function $h : \Sigma_\$ \mapsto \Sigma^*$, $h^*(\epsilon) = \epsilon$ and $h^*(a\omega) = h(a)h^*(\omega)$ for all $a \in \Sigma_\$$ and $\omega \in \Sigma_\$^*$. For example, when working with binary sequences, we usually encode "0" by "00", "1" by "11", and \$ by "01."

Let $A \subseteq \Sigma^*$ and $\nu : \mathbb{N} \to \Sigma^*$. Define the set $A_\nu = \{\omega\$\nu(|\omega|) \mid \omega \in A\}$, $A_\nu \subset \Sigma_\$^*$. Note that all words of A_ν that have the same length also receive the same suffix $\nu(|\omega|)$. This suffix is called the *advice*. We next encode A_ν back to Σ^* using a one to one homomorphism h^* as described above. We denote the resulting words by $\langle \omega, \nu(|\omega|) \rangle \in \Sigma^*$.

Remark 1.8.1 This encoding technique also allows us to explain how a discrete function f can be seen as a language $L(f)$. The language consists of the words $\langle \omega, \alpha(\omega) \rangle$: $\alpha(\omega) = f(\omega)$ if $f(\omega)$ is defined, otherwise $\alpha(\omega) = \epsilon$. □

The formal definition of nonuniform classes is as follows.

Definition 1.8.2 Nonuniformity: Given a class of languages C and a class of bounding functions H, we say that $A \in C/H$ if and only if there is a function $\nu \in H$ such that $h^*(A_\nu) \in C$.

Common choices for H are the space classes *poly* and *log*.

A special case is that of a Turing machine that receives polynomially long advice and computes in polynomial time. The class obtained in this fashion is called P/poly. When exponential advice is allowed, *any* language ranging to $\{0, 1\}$ is computable. Every such language L is decided by fixing the advice for an input of length n to have the bit "1" in location i ($i = 1, \ldots, 2^n$) if the ith word of length n (using the lexicographical order) is in L. This bit will be "0" otherwise.

In this book we concentrate on the special case of *prefix nonuniform classes* [BHM92]. In these classes, the advice $\nu(n)$ must be useful for all input strings of length up to n, not only those of length n. This is similar to the definitions of "strong" [Ko87] or "full" [BHM92] nonuniform classes. Furthermore, $\nu(n_1)$ is the prefix of $\nu(n_2)$ for all lengths $n_1 < n_2$. Formally, we define $\tilde{A}_\nu$ to be the set $\{\omega_1 \$ \nu(|\omega|) \mid \omega, \omega_1 \in A \text{ and } |\omega_1| < |\omega|\}$.

Definition 1.8.3 Prefix nonuniformity: Given a class of languages C and a class of bounding functions H, we say that $A \in \text{Pref-}C/H$ if and only if there is a prefix function $\nu \in H$ such that $h^*(\tilde{A}_\nu) \in C$. For the sake of brevity, we use the notation of C/H^* for the prefix advice class.

It has been proven that in many cases P/H^* coincides with P/H; for example $P/\text{poly}_* = P/\text{poly}$. Equality does not hold however for weaker function classes such as $P/\log$ or $\text{BPP}/\log$.

1.8.1 Circuit Families

There is an alternative way to define nonuniform classes, through families of Boolean circuits. The two definitions will be used interchangeably in Chapter 4. A *Boolean circuit* is a directed acyclic graph. Its nodes of in-degree 0 are called *input nodes*, while the rest are called *gates* which are labeled by one of the Boolean functions AND, OR, or NOT (the first two seen as functions of many variables, the last as a unary function). The constants "0" and "1" are produced by the Boolean gates (see Note 3) or can be implicitly added to the structure. In what follows we concentrate on languages rather than functions; as a consequence, one *output* node suffices. Adding extra gates if necessary, we can assume that nodes are arranged into levels $0, 1, \ldots, d$, where the input nodes (without incoming edges) are at level zero and the output node (with no outgoing edges) is at level d. Each Boolean gate has incoming edges only from the previous level, and the value it computes is used as an input by the next level. In this fashion each circuit computes a Boolean function of the inputs. The *depth* of the circuit is the number of levels d, its *width* is the maximum number of gates in any level, and its *size* is the total number of gates.

A *family of circuits* $\mathcal{C}$ is an indexed set

$$\{\mathcal{C}_n \mid n \in \mathbb{N}\},$$

where the circuit $\mathcal{C}_n$ computes on inputs of length n. The nth circuit is characterized by its size $S_{\mathcal{C}}(n)$, depth $D_{\mathcal{C}}(n)$, and width $W_{\mathcal{C}}(n)$. If $L \subseteq \{0,1\}^+$, we say that the language L is *computed by the family* $\mathcal{C}$ if the characteristic function of

$$L \bigcap \{0,1\}^n$$

is computed by $\mathcal{C}_n$ for each $n \in \mathbb{N}$.

The qualifier "nonuniform" reminds us that there is no requirement that circuit families be recursively described. The notions of weak/strong nonuniformity as well as prefix nonuniformity can be formulated in this model as well.

When the size of the circuits is bounded by a polynomial function (regardless of the width or depth), they compute the class P/poly. In exponential size circuits, all languages are recognizable.

1.8.2   Neural Networks and Advice Machines

The main contribution of this book is to introduce a particular type of neural network, to analyze it computationally, and suggest it as a standard for analog computation. This network has a fixed structure, corresponding to an unchanging number of "neurons." If allowed exponential time for computation, it turns out that the network computes all languages. Although this framework is more powerful than Turing machines, under polynomial-time constraints there are limits on its capabilities. In particular, there is a precise correspondence between networks and standard advice Turing machines with equivalent time resources, and as a consequence, a lower bound on the computability of networks can be determined. This relationship is perhaps surprising since our neural network is uniform and does not change in any manner with input size.

We show in Chapter 4 that networks with real weights compute P/poly; the stochastic model with rational weights and real probabilities computes only BPP/ $\log^*$ (Chapter 9). These two models are super-Turing, though the first is strictly stronger than the second.

We are now ready to proceed to the next chapter, in which this formal model is introduced.

1.9 Notes

1. This is not the case if one knows in advance that the strings to be memorized will be of no more than a certain predetermined length. In this case enough memory, represented by a certain number of states, can be preallocated for storage.

2. For simplicity we concentrate on a particular abstraction of the halting problem: is the predicate $\rho_{\tau(M)}$ total? Let g be the characteristic function of "$\rho_{\tau(M)}$ is total." We show that g is not computable. Define by f the function

$$f(\omega) = \begin{cases} \rho_\omega(\omega) + 1 & \text{if } g(\omega) = 1 \\ 0 & \text{if } g(\omega) = 0. \end{cases}$$

The function f is total and it is different from all computable functions. Furthermore, f can be written as

$$f(\omega) = \begin{cases} \rho_U(\omega, \omega) + 1 & \text{if } g(\omega) = 1 \\ 0 & \text{if } g(\omega) = 0. \end{cases}$$

If g were computable then f would also be computable, which is a contradiction. Thus g is not computable.

3. The constant $0(1)$ is obtained by computing the gate AND(OR) on any input and its complement (NOT).

Chapter 2

The Model

In this chapter we introduce the formal model of the neural network to be utilized and analyzed in this book.

Recurrent neural networks are composed of N elementary processors, called *neurons*. The ith processor is associated with an instantaneous *activation value*, or *local state* $x_i(t)$, which is a component of the activation vector x. At each time step a vector u of external binary inputs with components u_j, $j = 1, \ldots, M$, is presented to the system. The dynamics of the network is defined by a map

$$\mathcal{F} : \mathbb{R}^N \times \{0, 1\}^M \to \mathbb{R}^N, \tag{2.1}$$

which reads component-wise as:

$$x_i(t + 1) = \sigma \left(\sum_{j=1}^{N} a_{ij}\, x_j(t) + \sum_{j=1}^{M} b_{ij}\, u_j(t) + c_i \right), \quad i = 1, \ldots, N. \tag{2.2}$$

The letter σ represents the nonlinear *activation function* (also called the *response function*). The argument of the activation function in Equation (2.2) is a function of the input of the neuron, called the *net function*. The numbers a_{ij}, b_{ij}, and c_i are called the *weights* of the network. As part of the description of the network, we single out a subset of ℓ processors $x_{i_1}, \ldots, x_{i_\ell}$; these are the ℓ *output processors* (or *channels*), used to communicate the outputs of the network to the environment.

The weights of the network assume arbitrary real values. This is analogous to real constants in physical systems, which can be either basic physical constants such as c the speed of light, h (Planck's constant), and e (the charge of the electron); or macroscopic properties of physical systems such as elastic constants, viscosity, friction coefficients, resistances, etc. Obviously, it may be impossible in any practical sense to design a network having pre-specified non-rational weights; infinite precision would be required. It is similarly impractical to assume that the updates can occur with infinite precision. (Leaving aside the problems arising from the uncertainty principle of quantum mechanics, noise alone would make the perfect implementation of any such system impossible.) This leads us to consider variants that do not rely on infinite precision, e.g. bounded precision neurons and stochastic networks (see next section).

Throughout this book (except for the last chapter), we view the neural network as an idealized mathematical model. As such, the precise real values do not constitute an unreasonable constraint. Because recurrent neural networks are defined with unbounded precision, one might think that infinite precision would be required to fully describe their computation. We show, however, that this is not the case. We prove in Chapter 4 that when σ has some continuous properties, throughout the first q steps of the computation only the $O(q)$ significant bits may influence the result. This property of being indifferent to the least significant bits, or synonymously, to "small" perturbations in the value of the neurons (where "small" is measured as a function of the computation time) can be interpreted as a *partial immunity* to external noise or to architectural imprecision (Chapter 10). Though it does not appear in digital computational models, this *robustness* is a desirable computational feature.

Once the weights, the subset of output processors, and the function σ have been specified, the behavior of the network is completely defined for a given input encoding.

2.1   Variants of the Network

To obtain a better understanding of the expressive power of our model and its relation to other models, we examine the effect of changing the various parameters. The weights, for example, may take arbitrary real values, or rather, be constrained to subsets of them. We will see in this chapter and in Chapters 3-5 how these restrictions give rise to neural models with lower computational power.

We also consider modifications in the activation function. In Subsection 1.3.1 we introduced two common binary activation functions, the signal function:

$$signal(x) = \begin{cases} 0 & x \leq 0 \\ 1 & x > 0 \end{cases} \tag{2.3}$$

and the Heaviside (threshold) function:

$$\mathcal{H}(x) = \begin{cases} 0 & x < 0 \\ 1 & x \geq 0. \end{cases} \tag{2.4}$$

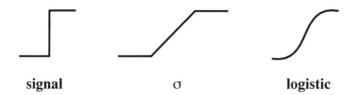

signal σ **logistic**

Figure 2.1: Different Activation Functions

Typical choices of continuous σ's are the logistic function $\frac{1}{1+e^{-x}}$ and the *piecewise linear* (*semilinear* or *saturated-linear*) function:

$$\sigma(x) = \begin{cases} 0 & \text{if } x < 0 \\ x & \text{if } 0 \le x \le 1 \\ 1 & \text{if } x > 1 . \end{cases} \tag{2.5}$$

Both of these are "sigmoidal," in the sense that they approximate the signal function when the "gain" γ is large in the expression $\sigma(\gamma x)$. Unless otherwise stated, we will take σ as the saturated-linear function. Changing the activation function results in different computational models, see Chapters 7 and 8.

In our model, the function σ is the same for all neurons. *Heterogeneous* models, in which different activation functions are allowed for different neurons, are also possible.

Another variant is obtained by using *high-order* neurons. Rather than being restricted to affine (i.e., first-order) combinations, as in Equation (2.2), the net functions of each high-order neuron is a polynomial in x and u:

$$x_i(t+1) = \sigma \left(\sum_{(|\alpha|+|\beta|) \le k} a_i^{\alpha\beta} x^\alpha(t) u^\beta(t) \right) \quad i = 1, \ldots, N, \tag{2.6}$$

where α, β are multi-indices,

$$x^\alpha = x_1^{\alpha_1} \ldots x_N^{\alpha_N} , \qquad u^\beta = u_1^{\beta_1} \ldots u_M^{\beta_M} ,$$

and "| |" denotes their magnitudes. The numbers $a_i^{\alpha\beta}$ are the weights of the network. High order and heterogeneous networks will be discussed in chapter 10.

One can also restrict the precision in the neurons, or add noise or stochasticity; these options are analyzed in Chapters 6 and 9, respectively. In addition to these variants, many others can be specified in analogy with other related models. In particular, we next demonstrate the relation of our model to the field of control.

2.1.1   A "System Diagram" Interpretation

As controlled dynamical systems (see [Son90]), networks can be viewed as discrete time systems built by combining delay lines with memory-free elements, each of which performs a nonlinear transformation on its input. For notational simplicity, we often write "$x^+(t)$" instead of "$x(t+1)$" (i.e., the superscript "$+$" indicates time shift) and drop the argument t. The update equation can therefore be written as:

$$x^+ = \sigma(Ax + Bu + c) \tag{2.7}$$

where c is a real N-vector and A, B are real matrices of sizes $N \times N$ and $N \times M$, respectively, and "σ" is interpreted as the application of σ to each coordinate of a vector argument. As for the vector c in Equation (2.7), observe that from the computational point of view, an equivalent model results by taking $c = 0$. To show the equivalence, add a local state x_{N+1} whose value is "hardwired" as "1", and let $a_{i,N+1} = c_i$ and $a_{N+1,N+1} = 1$. However, it will be very useful to leave c explicit, since this has the advantage of allowing us to take initial states to be $x = 0$, which corresponds to the intuitive idea that the system is at rest before the first input appears.

Although not essential for understanding any of the material to follow, it is worth presenting the model in the language of control theory. For that purpose, assume that the biases c_i are all zero, and introduce a matrix C so that the output can be thought of as $y = Cx$ (where C chooses the ℓ output neurons). Finally, denote by the symbol Δ the time-shift operator

$$\Delta(x(t)) = x^+(t)$$

on vector functions. Then the model being considered has the "system diagram" shown in Figure 2.2.

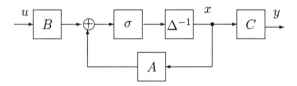

Figure 2.2: System Diagram for Networks.

In the special case in which σ is the identity, this is precisely a (discrete-time) linear control system, in the sense typically used in engineering. In this work, of course, we are primarily interested in nonlinear σ's, but the formal connection is nonetheless suggestive. One might expect that intuition from linear systems theory might help in deriving results about neural networks, and indeed, recent work shows this to be the case *.

2.2 The Network's Computation

We explore the computational power of our networks by studying the languages and functions that can be described as their I/O map. To clarify what we mean by a network recognizing a language

$$L \subseteq \{0, 1\}^+$$

we focus on networks whose output neurons take binary values only, and which comply with the I/O protocol that follows; these constitute the *formal networks* studied in this book. In this protocol, the input arrives on two binary input lines. The first of these is a *data line*, which is used to carry a binary input signal; when no signal is present, it defaults to zero. The second is the *validation line*, and it indicates when the data line is active. It takes the value "1" while the input is present and "0" thereafter. We use "D" and "V" to denote the contents of these two lines, respectively, so

$$u(t) = (D(t), V(t)) \in \{0, 1\}^2$$

for each t. Similarly, there are also two output processors that take the role of data and validation lines; they are denoted $G(t)$ and $H(t)$, respectively.

The convention of using two input lines allows all external signals to be binary. (Of course, many other conventions are possible and would give rise to the same results. For instance, one could use a three-valued input, say with values $\{-1, 0, 1\}$, where "0" indicates that no signal is present, and "± 1" are the two possible binary input values.)

We now encode each input string

$$\omega = \omega_1 \cdots \omega_k \in \{0, 1\}^+$$

by

$$u_\omega(t) = (D_\omega(t), V_\omega(t)) , \quad t \in \mathbb{N},$$

where

$$D_\omega(t) = \begin{cases} \omega_k & \text{if } t = 1, \ldots, k \\ 0 & \text{otherwise} \end{cases}$$

---

*See for instance [AS93b, AS93a] for work on identifying the weights from input/output data, [AS94] for results regarding the observability of activation values from outputs, and for reachability problems (characterizing the possible activations which may result after applying inputs).

and
$$V_\omega(t) = \begin{cases} 1 & \text{if } t = 1, \ldots, k \\ 0 & \text{otherwise .} \end{cases}$$

A word $\omega \in \{0,1\}^+$ is *classified in time* r by a formal net starting from the initial state $x(1) = 0$, if the input lines (D_ω, V_ω) take the values $D_\omega = \omega 0^\infty$ and $V_\omega = 1^{|\omega|} 0^\infty$, and the output line component $H_\omega(t) = 0$ for $t < r$ and $H_\omega(r) = 1$. If $G_\omega(r)$ is "1" then the word is accepted, and if $G_\omega(r)$ is "0" the word is rejected.

Definition 2.2.1 A language $L \subseteq \{0,1\}^+$ is *accepted* by a formal net $\mathcal{N}$ if, for every word $\omega \in L$, ω is accepted by $\mathcal{N}$, and for every word $\omega \notin L$, ω is rejected or not classified by $\mathcal{N}$. L is *recognized* or *decided* by a formal net $\mathcal{N}$ if L is accepted by $\mathcal{N}$ and its complement is rejected by $\mathcal{N}$. Let $T : \mathbb{N} \to \mathbb{N}$ be a total function on natural numbers. The language L is *recognized (decided) in time* T by the formal network $\mathcal{N}$ if any word $\omega \in \{0,1\}^+$ is correctly classified in time not greater than $T(|\omega|)$.

We similarly define networks that compute functions.

Definition 2.2.2 Let $\psi : \{0,1\}^+ \to \{0,1\}^+$ be a partial function and let $\mathcal{N}$ be a formal net with input lines (D, V) and output lines (G, H). The function ψ is *computable by the net* $\mathcal{N}$ if for every word $\omega \in \{0,1\}^+$:

1. if $\psi(\omega)$ is undefined, then $H_\omega = 0^\infty$

2. if $\psi(\omega)$ is defined, then there exists $r \in \mathbb{N}$ (called the *response time*) such that
$$G_\omega(t) = \begin{cases} \psi(\omega)[t - r + 1] & \text{if } t = r, \ldots, (r + |\psi(\omega)| - 1) \\ 0 & \text{otherwise} \end{cases}$$
and
$$H_\omega(t) = \begin{cases} 1 & \text{if } t = r, \ldots, (r + |\psi(\omega)| - 1) \\ 0 & \text{otherwise.} \end{cases}$$

The function ψ is *computable in time* $T : \mathbb{N} \to \mathbb{N}$ if for every $n \in \mathbb{N}$, $T(n)$ is defined if and only if ψ is defined on all inputs of size n, and there exists a formal network $\mathcal{N}$ such that, for every word $\omega \in \{0,1\}^+$, if $\psi(\omega)$ is defined the response time is at most $T(|\omega|)$).

Throughout this book we prove results on languages. We consider functions only when our framework "coincides" with (or "degenerates" to) the Turing model. This is done to emphasize that the classical theory of computation can be captured within our model.

Because the neurons may assume real values, finite networks are computationally powerful. Applying different constraints to the set of admissible weight values results in networks equivalent to various computational models. We close this chapter with the simple degenerate example of integer weights.

2.3   Integer Weights

In 1956, Kleene showed how to simulate finite automata using McCulloch and Pitts neurons [Kle56]. We, however, do not use Boolean neurons, but rather neurons that may take on analog values in $[0, 1]$, and as a result our networks are potentially more powerful. When the weights are constrained to be integers, the analog neurons are constrained to binary activations, and the networks reduce to become computationally equivalent to those studied by McCulloch and Pitts. Although this fact is fairly straightforward and essentially well-known, its inclusion here remains interesting for completeness. Furthermore, we use the rest of this section as a simple demonstration for the approach we take later on.

Theorem 1 *The languages accepted by integer networks are exactly the regular ones over the alphabet $\{0, 1\}$.*

Proof. In our model of computation (i.e., neural networks), decisions are indicated by a special output signal. To establish the desired equivalence, we need to define an *offline finite automaton*. This is an automaton that may continue computing after completely reading the input string, and decide on the input after some delay. A decision is reached when the computation arrives at either an accepting or a rejecting state.

We denote an offline finite automaton as a 5-tuple $(Q, \Sigma, f_{\text{off}}, q_0, F_{\text{off}})$. The transition function f_{off} maps $Q \times (\Sigma \bigcup \{\$\})$ into Q, where "$\$$" is a special symbol ($\$ \notin \Sigma$) denoting that the complete input string has been read into the machine. Here, $F_{\text{off}} \subseteq Q$ is the set of accepting states, and the transition function is such that $f_{\text{off}}(q, a) = q$, for all $a \in \Sigma \bigcup \{\$\}$ and $q \in F_{\text{off}}$. The objects Q, Σ, and q_0 are defined as in the finite automaton case. We extend the transition function f_{off} to be defined on a state and a string of the type $\omega\$^*$, where $\omega \in \Sigma^*$ and $\$^*$ is a string of 0 or more appearances of the "$\$$" sign. We define a function $\tilde{f}_{\text{off}}$ which maps $Q \times \Sigma^*$ (without the $\$$ sign) into 2^Q (the set of subsets of Q) by

$$\tilde{f}_{\text{off}}(q, \omega) = \{f_{\text{off}}^*(q, \omega\$^*)\}$$

for any $\omega \in \Sigma^*$. A string $\omega \in \Sigma^*$ is said to be *accepted* by an offline finite automaton $\mathcal{M}$ if $\tilde{f}_{\text{off}}(q_0, \omega) \bigcap F_{\text{off}} \neq \phi$.

Lemma 2.3.1 *The class $\mathcal{L}$ of languages accepted by offline finite automata is exactly the class of regular languages.*

Proof. Any finite automaton can be seen as an offline automaton by letting $f_{\text{off}}(q, \$) = q$ for all $q \in Q$. Thus, the regular languages are included in $\mathcal{L}$. To prove the other inclusion, we show next that given an offline finite automaton $\mathcal{M} = (Q, \Sigma, f_{\text{off}}, q_0, F_{\text{off}})$ that accepts a language L, there is a finite automaton $\mathcal{M}'$ that accepts the same language. Define $F = \{q \in Q \mid \tilde{f}_{\text{off}}(q, \$^*) \cap F_{\text{off}} \neq \phi\}$, and let f be the map f_{off} restricted to $Q \times \Sigma$. Then, the machine $\mathcal{M} = (Q, \Sigma, f, q_0, F)$ accepts L. ∎

We now establish a correspondence between integer networks and offline finite automata, required for the proof of Theorem 1. We show each inclusion separately.

1. Given a formal integer network $\mathcal{N}$ that consists of N neurons and accepts the language L, we define an offline finite automaton $\mathcal{M} = (Q, \{0, 1\}, f_{\text{off}}, q_0, F_{\text{off}})$ as follows:

   (a) We identify the input $(D(t), V(t))$ to $\mathcal{N}$ with the values in $\{0, 1, \$\}^2$ using the following encoding: $\zeta[(0, 1)] = 0$, $\zeta[(1, 1)] = 1$ and $\zeta[(0, 0)] = \$$ (the case $(1, 0)$ is invalid).

   (b) Let $Q = \{0, 1\}^N$ and $q_0 = 0^N$. (Note that as the network starts from an initial state 0^N and utilizes only integer weights, its neurons may assume binary values only.) That is, we identify each state of $\mathcal{M}$ with the activations of all the neurons.

   (c) Assume, without loss of generality, that $G(t)$ and $H(t)$ are the first and second neurons in the state encoding, respectively. We denote by $q[i]$ the ith coordinate of state q. Then, $F_{\text{off}} = \{q \in Q \mid q[1] = 1, q[2] = 1\}$.

   (d) The transition function f_{off} is induced by the update equation $q^+ = \sigma(Aq + Bu + c)$.

   It is easy to verify that $L(\mathcal{M}) = L$.

2. Given an offline finite automaton $\mathcal{M} = (Q, \Sigma, f_{\text{off}}, q_0, F_{\text{off}})$ that accepts a language $L \subseteq \{0, 1\}^*$, assume without loss of generality that there is no transition into the initial state q_0. We define a formal integer network $\mathcal{N}$ and the simulation of $\mathcal{M}$ as follows:

   The input letters from $\{0, 1, \$\}$ are translated via the function ζ^{-1}. The number of neurons is $N = 3|Q| + 2$. Of these, $3|Q|$ are indexed by $j = 0, \ldots, (|Q| - 1)$, and a pair (k, l) which may assume the values $(0, 0)$, $(0, 1)$, or $(1, 1)$. Each x_{jkl} may assume a binary value only, where x_{jkl} is 1 if and only if the current state of the machine $\mathcal{M}$ is q_j and its last input was $\zeta[(k, l)]$. The construction will be such that at each moment exactly one of these neurons has the value 1 and all the rest have the value 0.

   Before we show the update equations of the neurons x_{jkl}, we introduce $|Q|$ auxiliary binary variables p_i $(i = 0, \ldots, |Q| - 1)$, each of them having the value 1 if the state of the machine $\mathcal{M}$ is to become q_i. These variables can be defined by the activation values of the neurons x_{jkl} as follows:

$$
\begin{aligned}
p_0 &= 1 - \sum_{j,k,l} x_{jkl} \\
p_i &= \sum_{j,k,l} a^i_{jkl} \, x_{jkl}
\end{aligned}
$$

where the sum is over all $j = 0, \ldots, |Q| - 1$, and pairs (k,l) as before, and the constants a^i_{jkl} are such that $a^i_{jkl} = 1$ when $f_{\text{off}}(q_j, \zeta[(k,l)]) = q_i$ and $a^i_{jkl} = 0$ otherwise.

Now we are ready to precisely state the update equation of the neurons x_{ikl}:

$$
\begin{aligned}
x^+_{i11} &= \sigma(p_i + V + D - 2) \\
x^+_{i10} &= \sigma(p_i + 2V - D - 2) \\
x^+_{i00} &= \sigma(p_i - V - D) \,.
\end{aligned}
$$

Two additional neurons constitute the validation and data output. Define $F_R = \{q \in Q \mid \tilde{f}_{\text{off}}(q, \$^*) \cap F_{\text{off}} = \phi\}$; the validation neuron updates by $x^+_v = \sigma(\sum_{j=0}^{|Q|-1} a_j \, x_{j00})$, where a_j is "1" when $q_j \in (F_{\text{off}} \bigcup F_R)$ and is "0" otherwise. The data neuron updates by a similar equation, where $a_j = 1$ when $q_j \in F_{\text{off}}$ and $a_j = 0$ otherwise.

The simulation implies that $\mathcal{M}$ and $\mathcal{N}$ accept the same language. Note that all of the weights of $\mathcal{N}$ are integers. ∎

Chapter 3

Networks with Rational Weights

The full neural network model with real weights will be analyzed in the next chapter. Here, we discuss the model with the weights constrained to the set of rationals. In contrast to the case described in the previous chapter, where we dealt only with integer weights, and each neuron could assume two values only, here a neuron can take on countably infinite different values. The analysis of networks with rational weights is a prerequisite for the proofs of the real weight model in the next chapter. It also sheds light on the role of different types of weights in determining the computational capabilities of the model.

A finite network with rational weights is finitely describable and can be efficiently simulated by a Turing machine. We show that the converse holds as well: Turing machines can be efficiently simulated by networks having rational weights. In particular, it is possible to specify a network that simulates a universal Turing machine in real time; this can be done by using fixed precision rational numbers as weights. We conclude that these two models are computationally equivalent. Nondeterministic Turing machines and nondeterministic networks are shown to be equivalent as well.

A related result was obtained by Pollack [Pol87]. Pollack argued that a certain recurrent net model, which he called a "neuring machine," is universal. Pollack's set-up consists of a finite number of high-order neurons of two different kinds, with identity and threshold responses. This model is different from ours in the following ways: (a) it allows for discontinuous functions, (b) it is heterogeneous, (c) it is high-order. Because of these differences, our model can be considered more relevant for analog computation. Since Pollack's proof, it has been an open question whether high-order connections are really necessary in order to achieve computational universality, and Pollack conjectured that they are. His assumption was, for a time, rather widely accepted by the neural network community, and was frequently cited as a motivation for the use of high-order networks in the application and design of neural networks (see, e.g., [CSSM89, Elm90, GMC+92, Pol87, SCLG91, WZ89]). We show that high-order neurons are not computationally superior to first-order neurons.

The equivalence with Turing machines has many implications regarding the *decidability*, or more generally, the complexity of questions about our networks. For instance, there is no computable limit on the running time of a neural network. (Clearly, there are particular networks that have well-specified limits.) As a result, one cannot emulate a neural network using fixed precision arithmetic, although it is shown in the next chapter that the required precision is linear in the computation time (partial immunity). Thus, our construction may be thought of as a negative result concerning noisy neural networks. Equivalently, one cannot *a priori* ignore the presence of even the slightest noise or roundoff error, because our construction is sensitive to both effects when computing for long durations.

Another derivative of the halting problem is that one cannot automatically assume that a neural network converges or enters a detectable oscillatory state within any reasonable time bound. In other words, determining whether a given neuron ever assumes the value "0" (or "close to 0") or whether the network of the form

$$x(t+1) = \sigma(Ax(t) + c)$$

ever reaches an equilibrium point is undecidable. This is in contrast with the particular recurrent networks suggested for content-addressable retrieval and optimization, such as the Hopfield network (see, e.g., [HT85]), where convergence is assured. More on related work can be found in Note 1.

The rest of this chapter is organized as follows. Section 3.1 states the results. Section 3.2 outlines the main proof. Details of the proof are provided in Sections 3.3–3.6. In Section 3.7, the "universal" network is described, and in Section 3.8, the nondeterministic version of the networks is introduced.

3.1   The Turing Equivalence Theorem

We focus on networks with rational weights. Because the initial state is rational, after iterating Equation (2.2) on binary inputs, the state vector $x(t)$ remains rational. The update equation is now a restriction of Equation (2.1) to

$$\mathcal{F} : \mathbb{Q}^N \times \{0,1\}^2 \to \mathbb{Q}^N .$$

Our main result is stated in the following theorem.

Theorem 2 *Let $\psi : \{0,1\}^+ \to \{0,1\}^+$ be a function computable by a p-stack Turing machine in time $T : \mathbb{N} \to \mathbb{N}$. Then there exists a rational formal network $\mathcal{N}$ that computes $\psi(\omega)$ in time $T(|\omega|) + O(|\omega|)$ for every $\omega \in \{0,1\}^+$ such that $\psi(\omega)$ is defined, and the network does not halt if $\psi(\omega)$ is undefined.*

As a particular case, a network that simulates a universal Turing machine does in fact exist. We refer to such a network as "universal," although it is not universal in the network domain but only in the Turing sense. Note that Theorem 2 states not only computational equivalence, but also equivalence in time complexity.

In the standard Turing model, the input is encoded into the tape or stack rather than arriving as an input stream. Analogously, we next consider networks that encode the input and output into the initial and final states, respectively. For a neural network without inputs, we may think of the dynamics map $\mathcal{F}$ as a map from $\mathbb{Q}^N$ to $\mathbb{Q}^N$. In this case, we denote by $\mathcal{F}^j$ the jth iterate of $\mathcal{F}$. Starting from an initial state $\xi^0 \in \mathbb{Q}^N$, the jth iterate is $\xi^j = \mathcal{F}^j(\xi^0)$. We now state that if $\psi : \{0,1\}^+ \to \{0,1\}^+$ is a Turing computable function, then there exists a rational neural network $\mathcal{N}$ without inputs and an encoding of data into the initial state of $\mathcal{N}$, such that $\psi(\omega)$ is undefined if and only if the third neuron has an activation value always equal to zero. The function $\psi(\omega)$ is defined if the value of the third neuron ever becomes equal to one, in which case the first neuron encodes the result.

The following encodings will be used throughout this book. They are particularly useful for decoding strings from neurons having continuous activation functions, so that the retrieval requires a constant amount of time per bit. The motivation for these encodings is postponed to Subsection 3.2.1, here we present the definitions only.

Let b be a natural number, and $\alpha = \alpha_1 \alpha_2 \cdots$ a finite or infinite sequence of natural numbers smaller than b. The *interpretation* of the sequence α in base b is the number

$$\alpha \mid_b \equiv \sum_{i=1}^{|\alpha|} \frac{\alpha_i}{b^i} \ .$$

We focus on two particular cases, and define the encoding functions by the following formulae:

$$\delta_2(\epsilon) = 0 \quad \delta_2(\alpha) = \alpha \Big|_2 = \sum_{i=1}^{|\alpha|} \frac{\alpha_i}{2^i} \ , \tag{3.1}$$

and

$$\delta_4(\epsilon) = 0 \quad \delta_4(\alpha) = \alpha \Big|_4 = \sum_{i=1}^{|\alpha|} \frac{2\alpha_i + 1}{4^i} \ . \tag{3.2}$$

Thus δ_2 is the standard binary expansion, with the usual technical problem that 01^∞ denotes the same number as 10^∞. On the other hand, δ_4 is injective, and its image is the *4-Cantor set*. The 4-Cantor set can be expressed as the union of two sets: $\tilde{\Delta}_4 = \Delta_4 \bigcup \bar{\Delta}_4$, where

$$\bar{\Delta}_4 = \{\sum_{i=1}^{k} \frac{\beta_i}{4^i} \;\; \beta_i \in \{1,3\}, \; k \geq 0\} \tag{3.3}$$

is the "finite component" of $\tilde{\Delta}_4$, where for $k = 0$, we interpret this sum as 0. The "infinite component" of $\tilde{\Delta}_4$ is

$$\Delta_4 = \left\{ \sum_{i=1}^{\infty} \frac{\beta_i}{4^i} \;\middle|\; \beta \in \{1,3\}^\infty \right\} . \tag{3.4}$$

The elements of $\bar{\Delta}_4$ are precisely the image of δ_4 for finite sequences, and Δ_4 is the range of this function when restricted to infinite strings.

We also need the definition of:

$$\delta_{\bar{p}}(\epsilon) = 0$$
$$\delta_{\bar{p}}(\alpha) = \alpha\Big|_{10p^2} = \sum_{i=1}^{|\alpha|} \frac{10p^2 - 1 + 4p(\alpha_i - 1)}{(10p^2)^i} \quad p > 1.$$

We are now ready to state the theorem.

Theorem 3 *If $\psi : \{0,1\}^+ \to \{0,1\}^+$ is computable by a p-stack Turing machine in time T, and if we let η denote either δ_4 or $\delta_{\bar{p}}$, then there exists a rational net $\mathcal{N}$, with initial state*

$$\xi^0(\omega) \;=\; (\eta(\omega), 1, 0, \ldots, 0) \in \mathbb{Q}^N \,,$$

such that:

(a) *If η is δ_4, then $\mathcal{N}$ computes ψ in time $R \in O(T)$, i.e., if $\psi(\omega)$ is undefined, then for all j, the third coordinate $\xi^j(\omega)[3]$ of the state after j steps is "0". If instead $\psi(\omega)$ is defined, then there exists $r \leq R(|\omega|)$ such that*

$$\xi^j(\omega)[3] \;\dot{=}\; 0 \,, \quad j = 0, \ldots, r-1,$$
$$\xi^r(\omega)[3] \;=\; 1 \,,$$

and $\xi^j(\omega)[1] = \delta_4(\psi(\omega))$. (This is a linear time simulation.)

(b) *Furthermore, if η is $\delta_{\bar{p}}$ then real-time simulation can be performed, i.e. $R = T$.*

The next few sections include the proofs of both theorems. We start by proving Theorem 3, and then obtain Theorem 2 as a corollary. As the details of the proof of Theorem 3 are very technical, we start by sketching the main steps of the proof (Section 3.2). The proof itself is organized into several steps. We first show how to construct a rational network $\mathcal{N}$ that simulates a given multi-stack Turing machine M in time $R = 4T$ (T is the computation time of the Turing machine); this is done in Sections 3.3 and 3.4. In Section 3.5, we modify the construction into a rational network that simulates a given multi-stack Turing machine with no slowdown in the computation. After Theorem 3 is proved, we show in Section 3.6 how to add inputs and outputs to a "neural network without input/output," thus obtaining Theorem 2 as a corollary and ending the proofs.

3.2 Highlights of the Proof

This section is aimed at highlighting the main part of the proof of Theorem 3. We start with a p-stack machine; this model is equivalent to a standard Turing machine for $p \geq 2$ ([HU79]). Formally, a p-stack machine consists of a finite control and p binary stacks, unbounded in length. Each stack is accessed by a read-write head, which is located at the top element of the stack. At the beginning of the computation, the binary input sequence is written on stack 1. At each step, the machine reads the top element of each stack as a symbol $a \in \{0, 1, \#\}$ (where $\#$ means that the stack is empty), checks the state of the control ($q \in \{1, 2, \ldots, |Q|\}$), and executes the following operations:

1. For each stack, one of the following manipulations is made:
   (a) Popping the top element.
   (b) Pushing an extra element on the top of the stack (either "0" or "1").
   (c) No change in the stack.

2. Changing the state of the control.

When the control reaches a special state, called the "halting state," the machine stops. Its output is defined as the binary sequence on stack 1. Thus, the I/O map, or function computed by a p-stack machine, is defined by the binary sequences on stack 1 before and after the computation.

3.2.1 Cantor-like Encoding of Stacks

Let $\alpha = \alpha_1 \alpha_2 \ldots$ be the stack from top to bottom. Assume we were to encode a stack's binary stream $\alpha = \alpha_1 \alpha_2 \ldots$ with δ_2 into the number $\alpha|_2$. This value

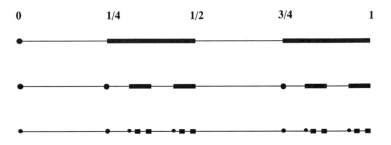

Figure 3.1: 4-Cantor set

could be held in a neuron since it ranges in $[0, 1]$; however it cannot be decoded efficiently in the sense that the number of operations required to retrieve the first bit is not fixed. Two different values that are close cannot be clearly differentiated with continuous activation functions. In order to distinguish between two close values it may be required to read *all* the bits representing the value. For example, in order to describe the first bit of the stacks $011 \cdots 1$ and $100 \cdots 0$, one must read the whole number. For infinite sequences this encoding is not well-defined, since it is not one-to-one.

We conclude that a good encoding should not range on a continuous interval but rather enforce gaps between valid encodings. Such gaps would enable a quick decoding of the number by means of an operation that requires only finite precision, or equivalently, that reads the most significant bit in some representation in a constant amount of time. On the other hand, if we choose some set of "numbers with gaps" to encode the different binary stacks, we have to assure that various manipulations on these numbers during the computation leave the stack encoding in the same set of "numbers with gaps."

As a solution, the stacks are encoded by δ_4:

$$g = \delta_4(\alpha) = \sum_{i=1}^{|\alpha|} \frac{2\alpha_i + 1}{4^i} \, .$$

The number g, as well as its suffixes when written in base 4, mav assume only a restricted set of values in $[0, 1)$ as is shown in Figure 3.1.

If the first digit to the right of the decimal point (i.e., the top of stack) is "0", then the value of the encoding ranges in $[\frac{1}{4}, \frac{1}{2})$; if it is "1", the value ranges in $[\frac{3}{4}, 1)$. The second digit after the decimal point decides the possible range relative to the current candidate range. In the above example of the stacks $011 \cdots 1$ and $100 \cdots 0$, their δ_4 encodings are $133 \cdots 3$ and $311 \cdots 1$, both interpreted in base 4. The operation of retrieving the first bit is translated into deciding whether the value is greater (or equal) than $\frac{3}{4}$ or smaller (or equal) than $\frac{1}{2}$. Because the value of the stacks never falls in the range $\frac{1}{2}$ and $\frac{3}{4}$, the decision is immediate.

The set of possible values is not continuous and has "holes." As demonstrated in the example above, the 4-Cantor set representation has the advantage that there is never a need to distinguish among two very close numbers in order to read the most significant bit (top of stack) out of the encoding; the top of stack can be retrieved in constant time employing a finite number of analog neurons.

3.2.2 Stack Operations

We next demonstrate the usefulness of our stack-encoding.

1. **Reading the Top:** Assume that a stack holds the value $\alpha = 1011$ from top to bottom that is encoded by the number $g = .3133_4$. As discussed above, the value of g is at least $\frac{3}{4}$ when the top of the stack is "1", and at most $\frac{1}{2}$ otherwise. The linear operation

$$4g - 2$$

transfers the range $[\frac{3}{4}, 1)$ that corresponds to the top element being "1" to $[1,2)$, and the range $[\frac{1}{4}, \frac{1}{2})$ to $[-1,0)$. Thus, the function

$$\text{top}(g) = \sigma(4g - 2)$$

saturates the resulting values into $\{0, 1\}$ and provides the value of the top element.

2. **Push:** Pushing "0" onto the stack $\alpha = 1011$ changes it into $\alpha = 01011$. In terms of the encoding, $g = .3133_4$ is transferred into $g = .13133_4$. That is, the suffix remains the same and the new element is entered into the most significant location. This is easily done by the operation

$$\frac{g}{4} + \frac{1}{4}$$

(which is equivalent to $\sigma(\frac{g}{4} + \frac{1}{4})$ given that $g \in [0,1)$.) Pushing the value "1" onto the stack can be implemented by $\frac{g}{4} + \frac{3}{4}$.

3. **Pop Stack:** Popping a stack transfers $\alpha = 1011$ into $\alpha = 011$, or the encoding from $g = .3133_4$ to $g = .133_4$. When the top element is known, the operation

$$4g - (2\text{top}(g) + 1)$$

(or equivalently $\sigma(4g - (2\text{top}(g)+1))$) has the effect of popping the stack.

4. **Non-empty Stack:** The predicate "non-empty" indicates whether the stack α is empty or not, which in terms of the encoding means whether $g = 0$ or $g \geq \frac{1}{4}$. This can be decided by the operation

$$\sigma(4g) \, .$$

3.2.3   General Construction of the Network

From the above discussion, we construct a network architecture that has three neurons per stack. One neuron holds the stack encoding g, one holds top(g), and one indicates whether or not the stack is empty. In addition, the architecture has several neurons that represent the finite control, and a set of neurons that take their input both from the stack reading neurons and from the finite control neurons and compute the next step.

Real-Time

The encoding suggested above, along with the use of an additional lemma (3.4.1) that describes a possible construction, results in a Turing machine simulation that requires four steps of the network for each step of the Turing machine. To achieve a "step by step" simulation, a more sophisticated encoding is required, which relies on a Cantor set with a specific gap-size. Then, we use large negative numbers as inhibitors, and in this manner attain the desired simulation in real-time.

Speed-Up

It is known that $\mathcal{P}$-stack machines ($\mathcal{P} \geq 2$) are polynomially equivalent to 2-stack machines ([HU79]). However, the simulation of a $\mathcal{P}$-stack machine by a 2-stack machine involves a slowdown that is more than linear. It is interesting to note that in our neural network, we are able to simulate any $\mathcal{P}$-stack machine (for any $\mathcal{P}$) in *real-time*. That is, for us, there is no slowdown in passing from one model to the other, and they are linear-time equivalent rather than polynomially equivalent.

We now turn to the proof itself.

3.3   The Simulation

As a departure point, we pick pushdown automata with $\mathcal{P}$ binary stacks. We choose to represent the values in the stacks as fractions with denominators that are powers of four. An algebraic formalization is as follows.

3.3.1 $\mathcal{P}$-Stack Machines

The instantaneous description of a p-stack machine, with a control unit of n states, can be represented by a $(p+1)$-tuple

$$(q, \delta_4(\omega_1), \delta_4(\omega_2), \ldots, \delta_4(\omega_p)),$$

where q is the state of the control unit, and the stacks store the words ω_h ($h = 1, \ldots, p$), respectively (later, in the simulation by a net, the state q will be represented by a vector).

Recall that $\bar{\Delta}_4$ is the "finite component" of the 4-Cantor set; it is the set of all rational numbers g that can be written in the form $g = \sum_{i=1}^{k} \frac{\beta_i}{4^i}$, $\beta_i \in \{1, 3\}$, $k \geq 0$. For any $g \in \bar{\Delta}_4$, we write

$$\zeta[g] = \begin{cases} 0 & \text{if } g \leq \frac{1}{2} \\ 1 & \text{if } g > \frac{1}{2}, \end{cases} \tag{3.5}$$

and:

$$\tau[g] = \begin{cases} 0 & \text{if } g = 0 \\ 1 & \text{if } g \neq 0. \end{cases} \tag{3.6}$$

We think of $\zeta[\cdot]$ as the "top of stack," because $\zeta[g] = 0$ when $\beta_1 = 1$ (and also when $g = 0$), and $\zeta[g] = 1$ when $\beta_1 = 3$. We interpret $\tau[\cdot]$ as the "non-empty stack" predicate. It can never happen that $\zeta[g] = 1$ while $\tau[g] = 0$; hence the pair $(\zeta[g], \tau[g])$ can have only three possible values in $\{0,1\}^2$.

A *p-stack machine* $\mathcal{M}$ is specified by a $(p+4)$-tuple

$$(Q, q_I, q_H, \theta_0, \theta_1, \theta_2, \ldots, \theta_p),$$

where Q is a finite set (of states), q_I and q_H are elements of Q called the *initial* and *halting states*, respectively, and the θ_i's are maps as follows:

$$\theta_0 \; : \; Q \times \{0,1\}^{2p} \to Q$$

$$\theta_h \; : \; Q \times \{0,1\}^{2p} \to \{(1,0,0), (\frac{1}{4}, 0, \frac{1}{4}), (\frac{1}{4}, 0, \frac{3}{4}), (4, -2, -1)\}$$

for $h = 1, \ldots, p$.

(The function θ_0 computes the next state, and the functions θ_i compute the next operation on stack h. The actions depend only on the state of the control unit and the symbol being read from each stack. The elements in the range

$$(1,0,0), (\frac{1}{4}, 0, \frac{1}{4}), (\frac{1}{4}, 0, \frac{3}{4}), (4, -2, -1)$$

of the θ_h should be interpreted as "no operation", "push0", "push1", and "pop", respectively.)

The set $\mathcal{X} = Q \times (\bar{\Delta}_4)^p$ is called the *instantaneous description set* of $\mathcal{M}$, and the map

$$\mathcal{P} : \mathcal{X} \to \mathcal{X}$$

defined by

$$
\begin{aligned}
\mathcal{P}(q, g_1, \ldots, g_p) = [\ & \theta_0(q, \zeta[g_1], \ldots, \zeta[g_p], \tau[g_1], \ldots, \tau[g_p]), \\
& \theta_1(q, \zeta[g_1], \ldots, \zeta[g_p], \tau[g_1], \ldots, \tau[g_p]) \cdot (g_1, \zeta[g_1], 1), \\
& \vdots \\
& \theta_p(q, \zeta[g_1], \ldots, \zeta[g_p], \tau[g_1], \ldots, \tau[g_p]) \cdot (g_p, \zeta[g_p], 1) \ ]
\end{aligned}
$$

where the dot "$\cdot$" indicates the inner product, is the *complete dynamics map* of $\mathcal{M}$. As part of the definition of $\mathcal{P}$, it is assumed that the maps θ_h, $h = 1, \ldots, p$, are such that $\theta_1(q, \zeta[g_1], \ldots, \zeta[g_p], 0, \tau[g_2], \ldots, \tau[g_p])$, $\theta_2(q, \zeta[g_1], \ldots, \zeta[g_p], \tau[g_1], 0, \tau[g_3], \ldots, \tau[g_p])$, $\ldots \neq (4, -2, -1)$, for all $q, g_1, \ldots, g_p$ (that is, one does not attempt to pop an empty stack).

Let $\omega \in \{0, 1\}^+$. Assume an initial configuration in which the control state is q_I, the first stack contains $\delta_4(\omega)$, and all other stacks are empty. If there exists a positive integer r, such that the machine reaches the halting state q_H after r steps, then the machine $\mathcal{M}$ is said to *halt on the input* ω. Let r be the smallest possible number such that

$$\mathcal{P}^r(q_I, \delta_4(\omega), 0, \ldots, 0) = (q_H, \delta_4(\omega_1), \delta_4(\omega_2), \ldots, \delta_4(\omega_p)) \ .$$

Then the machine $\mathcal{M}$ is said to *output the string* ω_1, and we let $\psi_{\mathcal{M}}(\omega) = \omega_1$. This defines a partial map

$$\psi_{\mathcal{M}} : \{0, 1\}^+ \to \{0, 1\}^+,$$

the I/O map of $\mathcal{M}$.

Except for the algebraic notation, the Turing computable functions $\psi : \{0, 1\}^+ \to \{0, 1\}^+$ are exactly the same as the maps $\psi_{\mathcal{M}} : \bar{\Delta}_4 \to \bar{\Delta}_4$ of p-stack machines as defined here; it is only necessary to identify words in $\{0, 1\}^+$ and elements of $\bar{\Delta}_4$ via the above encoding map δ_4.

3.4   Network with Four Layers

To prove the recursive power of neural networks, we simulate a p-stack machine, $\mathcal{M}$. Without loss of generality, we assume that the initial state q_I differs from the halting state q_H (otherwise the function computed is the identity, which can be easily implemented by a network). We identify the set of states $Q = \{q_1, \ldots, q_s\}$ with the set $\{1, 2, \ldots, s\}$, with $q_I = 1$, $q_H = 2$. We interpret the value 0 as a dummy state, which represents an inactive configuration of the machine prior to the beginning of the simulation.

We build the network in two stages.

Stage 1

As an intermediate step in the construction, we shall show how to simulate $\mathcal{M}$ with a certain dynamical system over $\mathbb{Q}^{s+p}$. We write a configuration as a vector in $\mathbb{Q}^{s+p}$.

$$(x_1, \ldots, x_s, g_1, \ldots, g_p) \, .$$

The first s components encode the state of the control unit; a state $i \in Q$ is encoded by the ith canonical vector

$$e_i = (0, \ldots, 0, 1, 0, \ldots, 0)$$

where the "1" is in the ith position. The g_h's encode the contents of the stacks. For notational ease, we substitute the top of stack $\zeta[g_h]$ and the non-empty stack predicate $\tau[g_h]$ by a_h and b_h, respectively. Formally, define

$$\beta_{ij} \; : \; \{0,1\}^{2p} \to \{0,1\}, \tag{3.7}$$

for $i \in \{1, \ldots, s\}$, $j \in \{1, \ldots, s\}$ and

$$\gamma_{hj}^k \; : \; \{0,1\}^{2p} \to \{0,1\}, \tag{3.8}$$

for $h \in \{1, \ldots, p\}$, $j \in \{1, \ldots, s\}$, $k = 1, 2, 3, 4$ as follows:

$$\beta_{ij}(a_1, a_2, \ldots, a_p, b_1, b_2, \ldots, b_p) = 1 \iff$$
$$\theta_0(j, a_1, a_2, \ldots, a_p, b_1, b_2, \ldots, b_p) = i$$

(intuitively: there is a transition from state j of the control part to state i if and only if the readings from the stacks are: top of stack h is a_h, and the non-emptiness test on stack h gives b_h). γ_{hj}^k is zero except in the following cases:

$$\gamma_{hj}^1(a_1, a_2, \ldots, a_p, b_1, b_2, \ldots, b_p) = 1 \iff$$
$$\theta_i(j, a_1, a_2, \ldots, a_p, b_1, b_2, \ldots, b_p) = (1, 0, 0)$$

(if the control is in state j and the stack readings are $a_1, \ldots, b_p$, then the stack h will not be changed),

$$\gamma_{hj}^2(a_1, a_2, \ldots, a_p, b_1, b_2, \ldots, b_p) = 1 \iff$$
$$\theta_i(j, a_1, a_2, \ldots, a_p, b_1, b_2, \ldots, b_p) = (\tfrac{1}{4}, 0, \tfrac{1}{4})$$

(if the control is in state j and the stack readings are $a_1, \ldots, b_p$, then the operation *push0* will occur on stack h),

$$\gamma_{hj}^3(a_1, a_2, \ldots, a_p, b_1, b_2, \ldots, b_p) = 1 \iff$$
$$\theta_i(j, a_1, a_2, \ldots, a_p, b_1, b_2, \ldots, b_p) = (\tfrac{1}{4}, 0, \tfrac{3}{4})$$

(if the control is in state j and the stack readings are $a_1, \ldots, b_p$, then the operation *push1* will occur on stack h),

$$\gamma_{hj}^4(a_1, a_2, \ldots, a_p, b_1, b_2, \ldots, b_p) = 1 \iff$$
$$\theta_i(j, a_1, a_2, \ldots, a_p, b_1, b_2, \ldots, b_p) = (4, -2, -1)$$

(if the control is in state j and the stack readings are $a_1, \ldots, b_p$, then the operation *pop* will occur on stack h).

Let $\widetilde{\mathcal{P}}$ be the map $\mathbb{Q}^{s+p} \to \mathbb{Q}^{s+p}$:

$$(x_1, \ldots, x_s, g_1, \ldots, g_p) \mapsto (x_1^+, \ldots, x_s^+, g_1^+, \ldots, g_p^+)$$

with

$$x_i^+ = \sum_{j=1}^{s} \beta_{ij}(a_1, \ldots, a_p, b_1, \ldots, b_p)\, x_j \tag{3.9}$$

for $i = 1, \ldots, s$ and

$$
\begin{aligned}
g_h^+ = \;& \left(\textstyle\sum_{j=1}^{s} \gamma_{hj}^1(a_1, \ldots, a_p, b_1, \ldots, b_p)x_j\right) g_h + & \text{(3.10.1)}\\
& \left(\textstyle\sum_{j=1}^{s} \gamma_{hj}^2(a_1, \ldots, a_p, b_1, \ldots, b_p)x_j\right)(\tfrac{1}{4}g_h + \tfrac{1}{4}) + & \text{(3.10.2)}\\
& \left(\textstyle\sum_{j=1}^{s} \gamma_{hj}^3(a_1, \ldots, a_p, b_1, \ldots, b_p)x_j\right)(\tfrac{1}{4}g_h + \tfrac{3}{4}) + & \text{(3.10.3)}\\
& \left(\textstyle\sum_{j=1}^{s} \gamma_{hj}^4(a_1, \ldots, a_p, b_1, \ldots, b_p)x_j\right)(4g_h - 2\zeta[g_h] - 1) & \text{(3.10.4)}
\end{aligned}
$$

$$\tag{3.10}$$

for $h = 1, \ldots, p$.

Let $\pi : \mathcal{X} \to \mathbb{Q}^{s+p}$ be defined by

$$\pi(i, g_1, \ldots, g_p) \;=\; (e_i, g_1, \ldots, g_p).$$

It follows immediately from the construction that

$$\widetilde{\mathcal{P}}(\pi(i, g_1, \ldots, g_p)) \;=\; \pi(\mathcal{P}(i, g_1, \ldots, g_p))$$

for all $(i, g_1, \ldots, g_p) \in \mathcal{X}$.

Applied inductively, the above implies that

$$\widetilde{\mathcal{P}}^r(e_1, \delta_4(\omega), 0, \ldots, 0) \;=\; \pi(\mathcal{P}^r(1, \delta_4(\omega), 0, \ldots, 0))$$

for all r, so $\psi(\omega)$ is defined if and only if for some r it holds that $\widetilde{\mathcal{P}}^r(e_1, \delta_4(\omega), 0, \ldots, 0)$ has the form

$$(e_2, g_1, \ldots, g_p)$$

(recall that for the original machine, $q_I = 1$ and $q_H = 2$, and these map respectively to e_1 and e_2 in the first s coordinates of the corresponding vector in $\mathbb{Q}^{s+p}$). If such a state is reached, then g_1 is in $\bar{\Delta}_4$ and its value is $\delta_4(\psi(\omega))$.

Stage 2

The second stage of the construction simulates the dynamics $\widetilde{\mathcal{P}}$ by a network. We first need the following technical fact.

Lemma 3.4.1 *Let $t \in \mathbb{N}$. For each function $\beta : \{0,1\}^t \to \{0,1\}$ there exist vectors*

$$v_1, v_2, \ldots, v_{2^t} \in \mathbb{Z}^{t+2}$$

and scalars

$$c_1, c_2, \ldots, c_{2^t} \in \mathbb{Z}$$

such that, for each $d_1, d_2, \ldots, d_t, x \in \{0,1\}$ and each $g \in [0,1)$,

$$\beta(d_1, d_2, \ldots, d_t)x = \sum_{r=1}^{2^t} c_r \sigma(v_r \cdot \mu) \tag{3.11}$$

and

$$\beta(d_1, d_2, \ldots, d_t)xg = \sigma\left(g + \sum_{r=1}^{2^t} c_r \sigma(v_r \cdot \mu) - 1\right), \tag{3.12}$$

where we denote $\mu = (1, d_1, d_2, \ldots, d_t, x)$ and "$\cdot$" denotes the inner product in $\mathbb{Z}^{t+2}$.

Proof. Write β as a polynomial

$$
\begin{aligned}
\beta(d_1, d_2, \ldots, d_t) &= c_1 + c_2 d_1 + \cdots + c_{t+1} d_t + c_{t+2} d_1 d_2 + \cdots \\
&\quad + c_{2^t} d_1 d_2 \cdots d_t,
\end{aligned} \tag{3.13}
$$

expand the product $\beta(d_1, d_2, \ldots, d_t)x$, and use that for any sequence $l_1, \ldots, l_k$ of elements in $\{0,1\}$,

$$l_1 \cdots l_k = \sigma(l_1 + \cdots + l_k - k + 1).$$

Use $x = \sigma(x)$ to get

$$
\begin{aligned}
\beta(d_1, d_2, \ldots, d_t)x &= \\
c_1 \sigma(x) + c_2 \sigma(d_1 + x - 1) &+ \cdots + c_{2^t} \sigma(d_1 + d_2 + \cdots + d_t + x - t) = \\
\sum_{r=1}^{2^t} c_r \sigma(v_r \cdot \mu) &
\end{aligned}
$$

for suitable c_r's and v_r's. On the other hand, for each $\tau \in \{0,1\}$ and each $g \in [0,1]$ it holds that $\tau g = \sigma(g + \tau - 1)$ (just check separately for $\tau = 0, 1$), so substituting τ in the above formula by $\tau = \beta(d_1, d_2, \ldots, d_t)x$ gives the desired result. ∎

Remark 3.4.2 The above construction overestimates the amount of neurons necessary. In the case where $t = 2p$ and the arguments are the top and non-empty functions of the stacks, the arguments are dependent, and there is a need for just 3^p terms in the summation, rather than 2^{2p}. This simplified version of $\beta(a_1, \ldots, a_p, b_1, \ldots, b_p)$ is obtained by imposing the constraints

$$a_i b_i = a_i \quad \text{for all } i = 1, \ldots, p. \tag{3.14}$$

After substituting Equation (3.14) in Equation (3.13), we reach a sum of terms, each of them with up to a degree p, made of non-complementary a_k and b_j (i.e., with $k \neq j$). For all degree i ($i = 0, \ldots, p$), we combine j ($j = 0, \ldots, i$) factors of the top of stack a_k (out of p stacks), with $(i - j)$ factors of the non-empty predicate b_k from the remaining $p - j$ stacks, to obtain

$$\sum_{i=0}^{p} \sum_{j=0}^{i} \binom{i}{j} \binom{p-j}{i-j} = 3^p$$

terms.   □

We now return to the proof of Theorem 3. Apply Lemma 3.4.1 repeatedly, with the "β" of the Lemma corresponding to each of the β_{ij}'s and γ_{hj}^k's, and using variously $g = g_h$, $g = (\frac{1}{4}g_h + \frac{1}{4})$, $g = (\frac{1}{4}g_h + \frac{3}{4})$, or $g = (4g_h - 2\zeta[g_h] - 1)$. Write also $\sigma(4g_h - 2)$ whenever $\zeta[g_h]$ appears, and $\sigma(4g)$ whenever $\tau[g]$ appears. The result is that $\widetilde{\mathcal{P}}$ can be written as a composition

$$\widetilde{\mathcal{P}} = F_1 \circ F_2 \circ F_3 \circ F_4$$

of four "saturated-affine" maps (i.e., maps of the form $\sigma(Ax + c)$): $F_4 : \mathbb{Q}^{s+p} \to \mathbb{Q}^\mu, F_3 : \mathbb{Q}^\mu \to \mathbb{Q}^\nu, F_2 : \mathbb{Q}^\nu \to \mathbb{Q}^\eta, F_1 : \mathbb{Q}^\eta \to \mathbb{Q}^{s+p}$, for some positive integers μ, ν, η. The argument to the function F_4, referred to below as z_1, of dimension $(s + p)$, represents the s x_i's of Equation (3.9) and the p g_h's of Equation (3.10). The functions F_1, F_2, F_3 compute the transition function of the x_i's and g_h's in three stages.

Consider the following set of equations:

$$z_1^+ = F_1(z_2)$$
$$z_2^+ = F_2(z_3)$$
$$z_3^+ = F_3(z_4)$$
$$z_4^+ = F_4(z_1) \,,$$

where the z_i's are vectors of sizes $s + p$, η, ν, and μ respectively. This set of equations models the dynamics of a σ-neural network, with

$$N = s + p + \mu + \nu + \eta$$

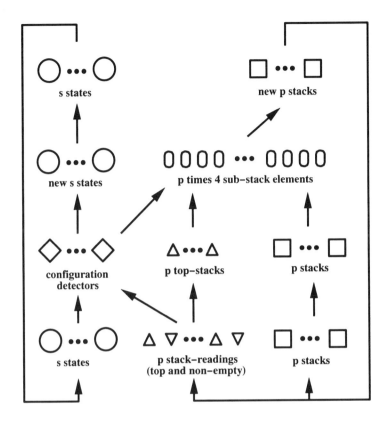

Figure 3.2: The Universal Neural Network

neurons. For an initial state of type $z_1 = (e_1, \delta_4(\omega), 0, \ldots, 0)$ and $z_i = 0$, $i = 2, 3, 4$, it follows that at each time of the form $t = 4k$, $k = 1, 2, \ldots$, the first block of coordinates, z_1, equals $\widehat{\mathcal{P}}^k(e_1, \delta_4(\omega), 0, \ldots, 0)$.

Reordering the coordinates so that the first stack $(s + 1)$st coordinate of z_1 becomes the first coordinate, and the previous $z_1[2]$ (that represented the halting state q_H of the machine $\mathcal{M}$) becomes the third coordinate, Theorem 3(a) is proved. ∎

3.4.1 A Layout Of The Construction

The above construction can be represented pictorially as in Figure 3.2.

It corresponds to the functions F_4, F_3, F_2, F_1, ordered from bottom to top. The neurons are divided into layers, where the output of the ith layer feeds into the $(i - 1)$st layer (and the output of the top layer feeds back into the bottom). The neurons are grouped in the picture according to their function.

The bottom layer, F_4, contains $p + 2$ groups of neurons. The leftmost group of neurons stores the values of the s states to pass to layer F_3. The "read stack h" group computes the top element $\zeta[g_h] \in \{0, 1\}$ and the non-empty predicate $\tau[g_h] \in \{0, 1\}$ of stack h. Each of the p neurons in the last group stores an encoding of one stack respectively.

Layer F_3 should compute the terms $\sigma(v_r \cdot \mu)$ that appear in Equation (3.11) for each of the possible s values of the vector x. We refer to these terms as "configuration detectors" because they provide the combinations of states and stack readings. Only $3^p s$ neurons are required, however, because there are only three possibilities for the ordered pair $(\zeta[g_h], \tau[g_h])$ for each stack; this was explained in Remark 3.4.2. Layer F_3 also preserves the values $\zeta[g_1], \ldots, \zeta[g_p]$ and the encoding of the p stacks from layer F_4.

Layer F_2 computes the new state as described in Equation (3.9) (to be passed along without change to the top layer). The four neurons in a "new stack h" group compute the four main terms (rows) of Equation (3.10) for stack h. For instance, for the fourth main term we compute an expression of the form:

$$\sigma(4g_h - 2\zeta[g_h] - 1 + \sum_{j=1}^{s} \sum_{r=1}^{2^{2p}} c_{rj}\sigma(v_r \cdot \mu_j) - 1)$$

(obtained by applying Equation (3.12)). Note that each of the terms g_h, $\zeta[g_h]$, $\sigma(v_r \cdot \mu)$ has been evaluated in the previous layer.

Finally, the top layer copies the states from layer F_3. It also adds the four main terms in each stack and applies σ to the result.

After reordering coordinates at the top layer to be

$$g_1, x_1, \ldots, x_s, g_2, \cdots, g_p \ ,$$

the data neuron and the halting neuron are first and third, respectively, as required to prove Theorem 3. Note that this construction results in values $\mu = s + 3p$, $\nu = 3^p s + 2p$, and $\eta = s + 4p$.

If desired, one could modify this construction so that all activation values are reset to zero when the halting neuron takes the value "1". This is discussed more thoroughly in Remark 3.6.1.

3.5   Real-Time Simulation

Here, we refine the simulation of Section 3.4 to obtain the claimed real-time simulation of Theorem 3(b). This proof is both less intuitive and less crucial for understanding our claims about Turing universality and complexity of neural networks.

We start in Subsection 3.5.1 by modifying the construction given in Section 3.4, in order to obtain a "two-layer" neural network. At this point, we obtain a slowdown of a factor of two in the simulation. In Subsection 3.5.2, we modify the construction so that in one of the layers, the neurons differ from the standard neurons: they compute linear combinations of their input with no sigma function applied to the combinations. This requires us to substitute the 4-Cantor set representation with a somewhat more complex Cantor set representation (to be explained there), which has larger gaps between consecutive admissible ranges of empty stacks, stacks with "0" top, and stacks with "1" top. These gaps enable the speed up of the simulation. Finally, in Subsection 3.5.3, we show how to modify the last network into a standard one with one layer only, thus achieving the desired real-time simulation of Turing machines.

3.5.1 Computing in Two Layers

We can rewrite the dynamics of the stack g_h from Equation (3.10) as the sum of four components:

$$g_h = \sum_{k=1}^{4} g_{hk} \, , \tag{3.15}$$

where g_{hk} represents the row (3.10.k) in Equation (3.10). We name these g_{hk} as the *sub-stack elements* of stack h. That is, g_{h1} may differ from "0" only if the last update of stack h was "no-operation", Similarly, the components g_{h2}, g_{h3}, g_{h4} may differ from "0" only if the last updates of the hth stack were "push0", "push1", or "pop", respectively. We also define t_{hk} to be the *sub-top* of the sub-stack element g_{hk}, and e_{hk} to the *sub-nonempty* test of the same sub-stack element. The top of stack h can be computed by

$$t_h = \sum_{k=1}^{4} t_{hk} \tag{3.16}$$

and

$$e_h = \sum_{k=1}^{4} e_{hk} \, . \tag{3.17}$$

As three out of the four sub-stack elements $\{g_{h1}, g_{h2}, g_{h3}, g_{h4}\}$ of each stack, $h = 1, \ldots, p$, are "0", and the fourth has the value of the stack g_h, it is also the case that three out of four sub-top elements of t_h (and sub-nonempty elements of e_h) are "0", and the fourth one stores the value of the top (nonempty predicate) of the relevant stack.

Using Equation (3.9), we can express the dynamics of the state control as

$$x_i^+ = \sum_{j=1}^s \beta_{ij}(t_{11}, \ldots, t_{p4}, e_{11}, \ldots, e_{p4})x_j \qquad (3.18)$$

for $i = 1, \ldots, s$; and the dynamics of the sub-stack elements as

$$g_{h1}^+ = \sigma(g_h + \sum_{j=1}^s \gamma_{hj}^1(t_{11}, \ldots, e_{p4})x_j - 1)$$

$$g_{h2}^+ = \sigma(\frac{1}{4}g_h + \frac{1}{4} + \sum_{j=1}^s \gamma_{hj}^2(t_{11}, \ldots, e_{p4})x_j - 1)$$

$$g_{h3}^+ = \sigma(\frac{1}{4}g_h + \frac{3}{4} + \sum_{j=1}^s \gamma_{hj}^3(t_{11}, \ldots, e_{p4})x_j - 1)$$

$$g_{h4}^+ = \sigma(4g_h - 2\zeta[g_h] - 1 + \sum_{j=1}^s \gamma_{hj}^4(t_{11}, \ldots, e_{p4})x_j - 1)$$

for all stacks g_h, $h = 1, \ldots, p$.

We introduce the notation

$$\text{next-}g_{hk} = \begin{cases} g_h & \text{if } k = 1 \\ \frac{1}{4}g_h + \frac{1}{4} & \text{if } k = 2 \\ \frac{1}{4}g_h + \frac{3}{4} & \text{if } k = 3 \\ 4g_h - 2\zeta[g_h] - 1 & \text{if } k = 4, \end{cases}$$

and summarize the above sub-stack dynamics equations by

$$g_{hk}^+ = \sigma\left(\text{next-}g_{hk} + \sum_{j=1}^s \gamma_{hj}^k(\cdot)x_j - 1\right). \qquad (3.19)$$

Similarly, the sub-top and sub-nonempty are updated by

$$t_{hk}^+ = \sigma\left(4\left[\text{next-}g_{hk} + \sum_{j=1}^s \gamma_{hj}^k(\cdot)x_j - 1\right] - 2\right), \qquad (3.20)$$

$$e_{hk}^+ = \sigma\left(4\left[\text{next-}g_{hk} + \sum_{j=1}^s \gamma_{hj}^k(\cdot)x_j - 1\right]\right).$$

(Note that each such μ contains $\zeta[g_1], \ldots, \zeta[g_p], \tau[g_1], \ldots, \tau[g_p]$, as well as $x_1, \ldots, x_s$, that were computed at layer F_4, for all $h = 1, \ldots, p$, $k = 1, \ldots, 4$.

We construct a network in which the stacks and their readings are not kept explicitly in values g_h, t_h, e_h, but implicitly only, via the sub-elements g_{hk}, t_{hk}, e_{hk} , $k = 1, \ldots, 4$, $h = 1, \ldots, p$. This will allow us to simulate one step of a Turing machine using a "two layer" rather than "four layer" network, as was suggested in the previous section.

By Lemma 3.4.1 and Equations (3.16), (3.17), the functions β_{ij} and γ_{hj}^k of Equations (3.18), (3.19) and (3.20) can be written as the combination

$$\sum_{j=1}^{s}\sum_{r=1}^{3^p} c_r^a \sigma(v_r^a \cdot \tilde{\mu}_j) \tag{3.21}$$

where

$$\tilde{\mu}_j = (1, \sum_{k=1}^{4} t_{1k}, \dots, \sum_{k=1}^{4} t_{pk}, \sum_{k=1}^{4} e_{1k}, \dots, \sum_{k=1}^{4} e_{pk}, x_j),$$

c_r^a are scalar constants, v_r^a are vector constants, and a represents the multi-indices ij for β and hjk for γ. Thus, all update equations of

$$\begin{aligned} x_i, \quad & i = 1, \dots, s & \text{(states)} \\ g_{hk} \quad & h = 1, \cdots, p, \quad k = 1, 2, 3, 4 \,, \\ t_{hk} \quad & h = 1, \cdots, p, \quad k = 1, 2, 3, 4 \,, \\ e_{hk} \quad & h = 1, \cdots, p, \quad k = 1, 2, 3, 4 \,, \end{aligned}$$

can be written as

$$\sigma(\text{ lin. comb. of } \sigma(\text{lin. comb. of } t_{hk} \text{ and } e_{hk}) \text{ and } g_{hk}),$$

that is, as what is usually called a "feedforward neural network with one hidden layer."

The main layer consists of the sub-elements g_{hk}, t_{hk}, e_{hk}, and the states x_i. In the hidden layer, we compute the $3^p s$ "configuration detectors" required by Lemma 3.4.1 to evaluate the functions β and γ, and thus the value of x_i^+. We also keep in the hidden layer the values of g_h and t_h to enable estimating next-g_{hk}. The result is that $\tilde{\mathcal{P}}$ can be written as a composition

$$\tilde{\mathcal{P}} = F_1 \circ F_2$$

of two "saturated-affine" maps: $F_1 : \mathbb{Q}^\nu \to \mathbb{Q}^\eta, F_2 : \mathbb{Q}^\eta \to \mathbb{Q}^\nu$, for $\nu = 3^p s + 2p$ and $\eta = s + 12p$.

In summary:

- The main layer consists of:

  1. s neurons x_i, $i = 1, \dots, s$, that represent the control state of the system unarily.

  2. For each stack h, $h = 1, \dots, p$, we have

     (a) four neurons $g_{hk}^1 = g_{hk}$, $k = 1, 2, 3, 4$,

(b) four neurons $t_{hk}^1 = t_{hk}$, $k = 1, 2, 3, 4$,

(c) four neurons $e_{hk}^1 = e_{hk}$, $k = 1, 2, 3, 4$.

- The hidden layer consists of:

  1. $3^p s$ neurons for configuration detecting.

  2. For each stack h, $h = 1, \ldots, p$, we have

     (a) a neuron $g_h^2 = g_h$,

     (b) a neuron $t_h^2 = t_h$.

3.5.2   Removing the Sigmoid From the Main Layer

Here, we proceed by shrinking the network and show how to construct an equivalent network to the one above, in which neurons in the main layer compute linear combinations only (and apply no σ function to it). In the following construction, we introduce a set of "noisy sub-stack" elements $\{\tilde{g}_{h1}, \tilde{g}_{h2}, \tilde{g}_{h3}, \tilde{g}_{h4}\}$, for each stack $h = 1, \ldots, p$. These may assume not only values in $[0, 1)$, but also negative values. Negative values of the stacks are interpreted as the value "0", whereas positive values are the true values of the stack. As in the last section, only one of these four elements may assume a non-negative value at each time. The "noisy sub-top" and "noisy sub-nonempty" functions applied to the noisy sub-stack elements may also produce values outside the range $[0, 1]$.

To manage with only one layer of σ functions, we need to choose a number representation that enforces large enough gaps between valid values of the stacks. We now abandon the 4-Cantor set representation, using instead a slightly different Cantor set representation: If the network is to simulate a p-stack machine, then our encoding is the $10p^2$-Cantor set.

To motivate our new Cantor representation, we illustrate it first with the special case $p = 2$. Here, the encoding is a 40-Cantor set representation. We encode the bit "0" by $\epsilon_0 = 31$ and "1" by $\epsilon_1 = 39$. We interpret the resulting sequence as a number in base 40. For example, the stack $\alpha = 1101$ is encoded by

$$g = .(39)(31)(39)(39)\big|_{40} \ .$$

In addition, we allow the empty stack to be represented by any non-positive value rather than by "0" only; this is where the sigmoids of the main layer are being saved. The possible ranges of a value g that represents a stack are thus:

$$
\begin{array}{ll}
(-\infty, 0] & \text{empty stack} \\
[\frac{31}{40}, \frac{32}{40}) & \text{top of stack is 0} \\
[\frac{39}{40}, 1) & \text{top of stack is 1} \ .
\end{array}
$$

Stack operations include push0, which is implemented by the operation $\frac{g}{40}+\frac{31}{40}$; push1, which is implemented as $\frac{g}{40}+\frac{39}{40}$; and pop, which is implemented by $(40g - 8t - 31)$, where t is the top element of the stack. We can compute the top and non-empty predicates by

$$
\begin{aligned}
\text{top}(g) &= 5(40g - 38) \,, \\
\text{non-empty } (g) &= (40g - 30) \,.
\end{aligned}
$$

The ranges of the top values are $[5, 10)$, $[-35, -30)$ and $(-\infty, -190]$ for top=1, top=0, and empty stack, respectively; and the ranges of the non-empty predicate are $[9, 10)$, $[1, 2)$, and $(-\infty, -30]$, in the same order. In particular, top is interpreted as being "1" when the function top(g) is in the range $[5, 10)$ and "0" in the range $(-\infty, -30]$, while the non-empty predicate is taken as "1" in the range $[1, 10)$ and "0" in the range $(-\infty, -30]$. The gaps between the positive and negative ranges are large; in particular, the absolute values that may represent the bit "0" in both predicates are at least 3 times larger than the values that represent the bit "1". The proposed encoding was designed to have this property. In the case of p-stack machines, we need an encoding for which the values of the negative domain are at least $(2p - 1)$ times larger than the values of the positive domain.

For the general p-stack machine, we choose the base $b = 10p^2$. Denote $c = 2p + 1$, $\epsilon_1 = (10p^2 - 1)$, and $\epsilon_0 = (10p^2 - 4p - 1)$. That is, "0" is encoded by ϵ_0, "1" is encoded by ϵ_1, and the resulting sequence is interpreted in base b. The role of c will be explained below. The reading functions "noisy sub-top" and "noisy sub-nonempty" that correspond to the noisy sub-stack elements $v \in \{\tilde{g}_{hk} \mid h = 1, \ldots, p, \ k = 1, \ldots, 4\}$ are defined as:

$$
\begin{aligned}
\text{N-top } (v) &= c(bv - (\epsilon_1 - 1)) & (3.22) \\
\text{N-nonempty } (v) &= bv - (\epsilon_0 - 1) \,. & (3.23)
\end{aligned}
$$

We denote $\tilde{t}_{hk} = $ N-top $(\tilde{g}_{hk})$ and $\tilde{e}_{hk} = $ N-nonempty $(\tilde{g}_{hk})$, for $h = 1, \ldots, p$, $k = 1, \ldots, 4$.

We summarize the dynamics equations of the noisy elements by

$$
\tilde{g}_{hk}^+ = \sigma\left(\text{next-}\tilde{g}_{hk} + \sum_{j=1}^{s} \gamma_{hj}^k(\cdot)x_j - 1\right) \,, \tag{3.24}
$$

$$
\tilde{t}_{hk}^+ = \sigma\left(c[b(\text{next-}\tilde{g}_{hk} + \sum_{j=1}^{s} \gamma_{hj}^k(\cdot)x_j - 1) - (\epsilon_1 - 1)]\right) \,,
$$

$$
\tilde{e}_{hk}^+ = \sigma\left(b(\text{next-}\tilde{g}_{hk} + \sum_{j=1}^{s} \gamma_{hj}^k(\cdot)x_j - 1) - (\epsilon_0 - 1)\right) \,,
$$

where

$$\text{next-}\tilde{g}_{hk} = \begin{cases} g_h & \text{if } k = 1 \\ \frac{1}{b}g_h + \frac{\epsilon_0}{b} & \text{if } k = 2 \\ \frac{1}{b}g_h + \frac{\epsilon_1}{b} & \text{if } k = 3 \\ bg_h - (\epsilon_1 - \epsilon_0)\zeta[g_h] - \epsilon_0 & \text{if } k = 4, \end{cases}$$

and $\zeta[g_h]$ is the true binary value of the top of stack h.

The ranges of values of the noisy sub-top and sub-nonempty functions are

$$\text{N-top } (v) \in \begin{cases} [2p + 1, 4p + 2) & \text{when the top is "1"} \\ [-8p^2 - 2p + 1, -8p^2 + 2) & \text{when the top is "0"} \\ (-\infty, -20p^3 - 10p^2 + 4p + 2] & \text{for an empty stack} \end{cases}$$

$$\text{N-nonempty } (v) \in \begin{cases} [4p + 1, 4p + 2) & \text{when the top is "1"} \\ [1, 2) & \text{when the top is "0"} \\ (-\infty, -10p^2 + 4p + 2] & \text{for an empty stack.} \end{cases}$$

(Note that the values of $\sigma(\text{N-nonempty}(v))$ and $\sigma(\text{N-top}(v))$ provide the exact binary top and non-empty predicates.) All of these intervals are within the range:

$$U = (-\infty, -8p^2 + 2) \cup [1, 4p + 2) \,.$$

The parameter c is chosen to insure large gaps in the range U:

Property: For all $p \geq 2$, any negative value of the functions N-top and N-nonempty has an absolute value of at least $(2p-1)$ times any positive value of them.

This is straightforward since $(2p - 1)$ times the upper limit of the range, i.e. $(4p + 2)$, gives $8p^2 - 2$ which is the absolute value of the upper limit of all possible negative values. The large negative numbers operate as inhibitors. We will see later how this property assists in constructing the network. As for the maintenance of negative values (that represent "0") in stack elements, Equation (3.15) is not valid anymore. That is, the elements $\tilde{g}_{hk}$, $k = 1, \ldots, 4$ cannot be combined linearly to provide the real value of the stack g_h. This is also the case with the top and non-empty predicates (see Equations (3.16) and (3.17)).

In Lemma 3.4.1, we proved that for any Boolean function of the type $\beta : \{0,1\}^t \mapsto \{0,1\}$ and $x \in \{0,1\}$, one may express

$$\beta(d_1, \ldots, d_t)x = \sum_{r=1}^{2^t} c_r \sigma(v_r \cdot \mu)$$

for $\mu = (1, d_1, \ldots, d_t, x)$, constants c_r and constant vectors v_r. This was applicable for the functions β_{ij} and γ_{hj}^k in Equations (3.18), (3.19), and (3.20).

Next, we prove that using the noisy sub-top $\tilde{t}_{hk}$ and noisy sub-nonempty $\tilde{e}_{hk}$ elements—rather than the binary sub-top $(\sigma(\tilde{t}_{hk}))$ and sub-nonempty $(\sigma(\tilde{e}_{hk}))$ ones—one may still compute the functions β_{ij} and γ_{hj}^k using one hidden layer only.

It is not true that for any function $\beta : U_t \to \{0,1\}$ and for any $x \in \{0,1\}$, $\beta(\cdot)x$ is computable in one hidden layer network; this is true in particular cases only, including ours.

Definition 3.5.1 A function $\beta(v_1,\ldots,v_t)$ is said to be *signal-invariant* if, for all $i = 1,\ldots,t$, $\beta(v_1,\ldots,v_t) = \beta(v_1',\ldots,v_t')$, for $v_i' = \text{signal}(v_i)$.

Lemma 3.5.2 *For each $t,r \in \mathbb{N}$, let U_t be the range*

$$(-\infty, -2t^2 + 2) \cup [1, 2t + 2),$$

and let

$$S_{r,t} = \{\, d \mid d = (d_1^{(1)}, \ldots, d_1^{(r)}, d_2^{(1)}, \ldots, d_2^{(r)}, \ldots, d_t^{(1)}, \ldots, d_t^{(r)}) \in U_t^{rt},$$

$$\text{and for all } i = 1, \ldots, t, \text{ at most one of the } d_i^{(k)} \text{ is positive} \,\}.$$

We denote by I the set of multi-indices $(i_1,\ldots,i_t)$, with $i_j \in \{0,1,\ldots,r\}$, for all $j = 1,\ldots,t$. For each function $\beta : S_{r,t} \to \{0,1\}$ that is signal invariant, there exist vectors

$$\{v_i \in \mathbb{Z}^{t+2} \mid i \in I\}$$

and scalars

$$\{c_i \in \mathbb{Z} \mid i \in I\}$$

such that, for each $(d_1^{(1)}, \ldots, d_t^{(r)}) \in S_{r,t}$ and any $x \in \{0,1\}$, we can write

$$\beta(d_1^{(1)}, \ldots, d_t^{(r)})x = \sum_{i \in I} c_i \sigma(v_i \cdot \mu_i),$$

where

$$\mu_i = \mu_{(i_1,\ldots,i_t)} = (1, d_1^{(i_1)}, d_2^{(i_2)}, \ldots, d_t^{(i_t)}, x), \qquad (3.25)$$

and where we are defining $d_i^{(0)} = 0$. Here the size of I is $|I| = (r+1)^t$, and "$\cdot$" denotes the inner product in $\mathbb{Z}^{t+2}$.

Proof. As β is signal-invariant, we can write β when acting on $S_{r,t}$ as

$$\beta(d_1^{(1)}, d_1^{(2)}, \ldots, d_t^{(r)}) = \beta(\sigma(d_1^{(1)}), \sigma(d_1^{(2)}), \ldots, \sigma(d_t^{(r)})) \,.$$

Thus, β can be viewed as a Boolean function from $\{0,1\}^{rt}$ to $\{0,1\}$, and we can express it as a polynomial (see Equation (3.13)):

$$\begin{aligned}
\beta(d_1^{(1)}, d_1^{(2)}, \ldots, d_t^{(r)}) = {}& c_1 + c_2\sigma(d_1^{(1)}) + c_3\sigma(d_1^{(2)}) + \cdots + c_{rt+1}\sigma(d_t^{(r)}) \\
& + c_{rt+2}\sigma(d_1^{(1)})\sigma(d_2^{(1)}) + \cdots \\
& + c_{(r+1)^t}\sigma(d_1^{(r)})\sigma(d_2^{(r)})\cdots\sigma(d_t^{(r)}) \,.
\end{aligned}$$

(Note that no term with more than t elements of the type $\sigma(d_k^{(l)})$ appears, as most $\sigma(d_k^{(l)}) = 0$, by definition of $S_{r,t}$.)   Observe that for any sequence $l_1, \ldots, l_k$ of $(k \leq t)$ elements in U_t and $x \in \{0, 1\}$, one has

$$\sigma(l_1) \cdots \sigma(l_k)x = \sigma(l_1 + \cdots + l_k + k(2t + 2)(x - 1)).$$

This is due to two facts:

1. The sum of k, $k \leq t$, elements of U_t is non-positive when at least one of the elements is negative. This stems from the property that any negative value in this range is at least $(t - 1)$ times larger than any positive value there.

2. Each l_i is bounded by $(2t + 2)$.

Expand the product $\beta(d_1^{(1)}, \ldots, d_t^{(r)})x$, using the above observation and the fact $x = \sigma(x)$. This leads to

$$
\begin{aligned}
\beta(d_1^{(1)}, \ldots, d_t^{(r)})x &= c_1\sigma(x) + c_2\sigma\left(d_1^{(1)} + (2t + 2)(x - 1)\right) + \cdots \\
&\quad + c_{(r+1)^t}\sigma\left(d_1^{(r)} + \cdots + d_t^{(r)} + t(2t + 2)(x - 1)\right) \\
&= \sum_{i \in I} c_i\sigma(v_i \cdot \mu_i) \ ,
\end{aligned}
$$

for suitable c_i's and v_i's, where μ_i is defined as in (3.25).   ∎

Remark 3.5.3 Note that in the case where the arguments are the functions N-top and N-nonempty, the arguments are dependent and not all $(r + 1)^t$ terms are needed.   □

We conclude that the noisy sub-stack elements $\tilde{g}_{hk}$, as well as the next control state, are computable in the one hidden layer network from $\tilde{t}_{hk}$ and $\tilde{e}_{hk}$.

We next provide the exact network architecture.

Network Description

The network consists of two layers. The main layer consists of both s state neurons and the neurons: $\tilde{g}_{hk}^1$, $\tilde{t}_{hk}^1$, $\tilde{e}_{hk}^1$, $h = 1, \ldots, p$, $k = 1, \ldots, 4$, that represent, respectively, noisy sub-stack elements, noisy sub-top elements, and noisy sub-nonempty elements.

For simplicity, we first provide the update equation of the network that simulates a 2-stack machine, and later generalize it to p-stack machines. Recall that for a 2-stack machine, the encoding was a 40-Cantor set representation with $\epsilon_0 = 31$ and $\epsilon_1 = 39$. We use the same notations for the functions γ_{hj}^k as in Equation (3.10); let g_h represent the hth stack, and t_h represents the true (i.e., clean) value of the top of the stack. The noisy sub-stack $\tilde{g}_{h1}$ updates for no-operation as in Equation (3.i10.1):

$$\tilde{g}_{h1}^{1+} = g_h + \sum_{j=1}^{s} \gamma_{hj}^1(\cdot)x_j - 1,$$

the noisy sub-stack $\tilde{g}_{h2}$ updates for push0 as in Equation (3.10.2):

$$\tilde{g}_{h2}^{1+} = \frac{g_h}{40} + \frac{31}{40} + \sum_{j=1}^{s} \gamma_{hj}^2(\cdot)x_j - 1,$$

the noisy sub-stack $\tilde{g}_{h3}$ updates for push1 as in Equation (3.10.3):

$$\tilde{g}_{h3}^{1+} = \frac{g_h}{40} + \frac{39}{40} + \sum_{j=1}^{s} \gamma_{hj}^3(\cdot)x_j - 1,$$

and the noisy sub-stack $\tilde{g}_{h4}$ updates for pop as in Equation (3.10.4):

$$\tilde{g}_{h4}^{1+} = 40g_h - 8t_h - 31 + \sum_{j=1}^{s} \gamma_{hj}^4(\cdot)x_j - 1,$$

where t_h denotes the top of stack h. For a general p-stack machine, the base is $10p^2$ and the update equations are given by:

$$\tilde{g}_{hj}^{1+} = \text{next-}\tilde{g}_{hj} + \sum_{j=1}^{s} \gamma_{hj}^k(\cdot)x_j - 1 \tag{3.26}$$

$$\tilde{t}_{hj}^{1+} = (2p+1)[10p^2(\text{next-}\tilde{g}_{hj} + \sum_{j=1}^{s} \gamma_{hj}^k(\cdot)x_j - 1) - (10p^2 - 2)] \tag{3.27}$$

$$\tilde{e}_{hj}^{1+} = 10p^2(\text{next-}\tilde{g}_{hj} + \sum_{j=1}^{s} \gamma_{hj}^k(\cdot)x_j - 1) - (10p^2 - 4p - 2) \tag{3.28}$$

where

$$\text{next-}\tilde{g}_{hk} = \begin{cases} g_h & \text{if } k = 1 \text{ (no update)} \\ \frac{1}{10p^2}g_h + \frac{10p^2 - 4p - 1}{10p^2} & \text{if } k = 2 \text{ (push0)} \\ \frac{1}{10p^2}g_h + \frac{10p^2 - 1}{10p^2} & \text{if } k = 3 \text{ (push1)} \\ 10p^2 g_h - 4pt_h - (10p^2 - 4p - 1) & \text{if } k = 4 \text{ (pop)} \end{cases}$$

and g_h and t_h are the exact values of the stacks and top elements. Using Lemma 3.5.2, all of the expressions of the type $\beta(\cdot)x$ and $\gamma(\cdot)x$ can be written as linear combinations of terms like

σ(linear combinations of $\tilde{g}^1_{hk}$, $\tilde{t}^1_{hk}$, $\tilde{e}^1_{hk}$).

These $\tilde{g}_{hk}$, $\tilde{t}_{hk}$, and $\tilde{e}_{hk}$ constitute the main layer.

The hidden layer consists of up to ($5^{2p}s$) configuration detector neurons (as proved in Lemma 3.5.2, having $r = 4$ and $t = 2p$) and the stack and top neurons:

$$g^2_{hk}, t^2_{hk}, \quad h = 1, \ldots, p, \quad k = 1, \ldots, 4,$$

which are updated by the equations

$$
\begin{aligned}
g^{2+}_{hk} &= \sigma(\tilde{g}^1_{hk}) & & [g_h = \textstyle\sum_{k=1}^4 g^2_{hk}] , \\
t^{2+}_{hj} &= \sigma(\tilde{t}^1_{hj}) , \quad h = 1, \ldots, p, \ k = 1, \ldots, 4 & & [t_h = \textstyle\sum_{k=1}^4 t^2_{hk}] .
\end{aligned}
$$

3.5.3   One Layer Network Simulates TM

Consider the above network. Remove the main layer and leave the hidden layer only, while letting each neuron there compute the information that it received beforehand from a neuron at the main layer. This can be written as a standard network and it ends the proof of Theorem 3.   ∎

3.6   Inputs and Outputs

We now explain how to deduce Theorem 2 from Theorem 3. In this section, we present details for the linear-time simulation (i.e., for the case of Theorem 3(a) and the encoding δ_4). The real-time case (Theorem 3(b) and $\delta_{\bar{p}}$) is entirely analogous, but the notations become more complicated.

We first show how to modify a network with no inputs into one which, given the input u_ω, produces the encoding $\delta_4(\omega)$ as a state coordinate and after that emulates the original net. Later we show how the output is decoded. As explained above, there are two input lines: $D = u_1$ carries the data, and $V = u_2$ validates it.

Assume we are given a network with no inputs

$$x^+ = \sigma(Ax + c) \tag{3.29}$$

as in the conclusion of Theorem 3. Suppose that we have already found a network

$$y^+ = \sigma(Fy + lu_1 + hu_2) \tag{3.30}$$

(consisting of 5 neurons) so that, if $u_1 = D_\omega$ and $u_2 = V_\omega$, then we have

$$y_4 = 00^{|\omega|+1}\delta_4(\omega)0^\infty \quad \text{and} \quad y_5 = 00^{|\omega|+2}1^\infty ,$$

that is,

$$y_4(t) = \begin{cases} \delta_4(\omega) & \text{if } t = |\omega| + 2 \\ 0 & \text{otherwise,} \end{cases} \quad \text{and} \quad y_5(t) = \begin{cases} 0 & \text{if } t \le |\omega| + 2 \\ 1 & \text{otherwise.} \end{cases}$$

Once this is done, modify the original network (3.29) as follows. The new state consists of the vector (x, y), with y evolving according to (3.30) and the equations for x modified in this manner (using A_i to denote the ith row of A and c_i for the ith entry of c):

$$\begin{aligned} x_1^+ &= \sigma(A_1 x + c_1 y_5 + y_4) \\ x_2^+ &= \sigma(A_2 x + c_2 y_5 + 4y_4) \\ x_i^+ &= \sigma(A_i x + c_i y_5), \quad i = 3, \dots, N. \end{aligned}$$

Then, starting at the initial state $y = x = 0$, clearly $x_1(t) = 0$ for $t = 0, \dots, |\omega| + 2$ and $x_1(|\omega| + 3) = \delta_4(\omega)$, $x_2(t) = 0$ for $t = 0, \dots, |\omega| + 2$, and $x_2(|\omega| + 3) = 1$, while, for $i > 2$, $x_i(t) = 0$ for $t = 0, \dots, |\omega| + 3$.

After time $|\omega| + 3$, as y_5 is "1" and u_1 and u_2 are "0", the equations for x evolve as in the original network, so that $x(t)$ in the new network equals $x(t - |\omega| - 3)$ in the original one for $t \ge |\omega| + 3$.

The system (3.30) can be constructed as follows:

$$\begin{aligned} y_1^+ &= \sigma(\frac{1}{4}y_1 + \frac{1}{2}u_1 + \frac{5}{4}u_2 - 1) \\ y_2^+ &= \sigma(u_2) \\ y_3^+ &= \sigma(y_2 - u_2) \\ y_4^+ &= \sigma(y_1 + y_2 - u_2 - 1) \\ y_5^+ &= \sigma(y_3 + y_5). \end{aligned}$$

This completes the proof of the encoding part. For the decoding process of producing the output signal, it will be sufficient to show how to build a network (of dimension 10 and with two inputs) such that, starting at the zero state and if the input sequences are x_1 and x_3, where $x_1(k) = \delta_4(\omega)$ for some k, and $x_3(t) = 0$ for $t < k$, $x_3(k) = 1$ ($x_1(t) \in [0, 1]$ for $t \ne k$, $x_3 \in [0, 1]$ for $t > k$), then for neurons z_9, z_{10} it holds that

$$z_9(t) = \begin{cases} 1 & \text{if } k + 4 \le t \le k + 3 + |\omega| \\ 0 & \text{otherwise}, \end{cases}$$

and

$$z_{10}(t) = \begin{cases} \omega_{t-k-3} & \text{if } k + 4 \le t \le k + 3 + |\omega| \\ 0 & \text{otherwise}. \end{cases}$$

One can verify that this can be done by:

$$
\begin{aligned}
z_1^+ &= \sigma(x_3 + z_1) \\
z_2^+ &= \sigma(z_1) \\
z_3^+ &= \sigma(z_2) \\
z_4^+ &= \sigma(x_1) \\
z_5^+ &= \sigma(z_4 + z_1 - z_2 - 1) \\
z_6^+ &= \sigma(4z_4 + z_1 - 2z_2 - 3) \\
z_7^+ &= \sigma(16z_8 - 8z_7 - 6z_3 + z_6) \\
z_8^+ &= \sigma(4z_8 - 2z_7 - z_3 + z_5) \\
z_9^+ &= \sigma(4z_8) \\
z_{10}^+ &= \sigma(z_7) \, .
\end{aligned}
$$

In this case the output is $y = (z_{10}, z_9)$.

Remark 3.6.1 If one would also like to achieve a resetting of the whole network after completing the operation, it is possible to add the neuron

$$
z_{11}^+ = \sigma(z_{10}) \, ,
$$

and to add to each neuron that is not identically zero at this point of time,

$$
v_i^+ = \sigma(\cdots + z_{11} - z_{10}) \, , \; v \in \{x, y, z\} \, ,
$$

where "$\cdots$" is the formerly defined operation of the neuron.   □

3.7   Universal Network

The number of neurons required to simulate a Turing machine consisting of s states and p stacks, with a slowdown of a factor of two in the computation, is:

$$
\underbrace{s + 12p}_{\text{main layer}} \; + \; \underbrace{3^p s + 2p}_{\text{hidden layer}} \quad .
$$

To estimate the number of neurons required for a "universal" neural network, we should calculate the number s discussed above, which is the number of states in the control unit of a two-stack universal Turing machine. Minsky proved the existence of a universal Turing machine having one tape with 4 letters and 7 control states [Min67]. Shannon showed how to control the number of letters and states in a Turing machine [Sha56]. Following his construction, we obtain a 2-letter 63-state 1-tape Turing machine. However, we are interested in a two-stack machine rather than one tape. Similar arguments to the ones made by Shannon, but for two stacks, lead us to $s = 84$. Applying the formula $3^p s + s + 14p$, we conclude that there is a universal network with 868 neurons. To allow for input and output to the network, we need an extra 16 neurons, thus having 884 in a "universal machine." (This estimate is very conservative. It is quite possible that with some care in the construction one may be able to reduce this estimate drastically. One useful tool may, for example, be the result in [ADO91] applied to the control unit.)

In this chapter we not only showed that first-order neural networks are universal, but that they are capable of simulating any Turing machine in real-time.Had our concern been solely to prove that neural networks are universal, then much simpler constructions, such as [NSCA97, SS91], would have sufficed. Smaller universal networks have been suggested in [Ind95, KCG94]; if high-order neurons are allowed, then a small network of nine neurons suffices (see [SM98]).

3.8 Nondeterministic Computation

A *nondeterministic neural network* is a modification of a deterministic network, obtained by incorporating a third constrained input line—called *guess line* and denoted by $\mathcal{G}$—in addition to the validation and data lines. For simplicity we restrict our attention to formal nondeterministic networks that compute binary output values only, which correspond to accept/reject.

Let $\omega \in \{0,1\}^+$, and denote by S_r the set $\{0,1\}^r$, and by γ_r an element in S_r. The dynamics of the nondeterministic network $\mathcal{N}$ for input ω is the result of a sequence of stages where, in stage $r = 1, 2, \ldots$, $\mathcal{N}$ acts on the inputs D_ω and V_ω, and on all possible strings $\mathcal{G} = \gamma_r$ (in parallel). For each possible value γ, the network has a dynamics map of the form

$$\mathcal{F}(x, u, \gamma) = \sigma(Ax + b_1 u_1 + b_2 u_2 + b_3 \gamma + c),$$

for some matrix $A \in \mathbb{Q}^{N \times N}$ and four vectors $b_1, b_2, b_3, c \in \mathbb{Q}^N$.

The rule for acceptance or rejection of an input string ω is the same as in the classical theory of computation. We say that ω is classified in time r by a nondeterministic formal network $\mathcal{N}$, if $\mathcal{N}$ either halts on input ω for *a* guess $\gamma_r \in S_r$ and announces acceptance, or $\mathcal{N}$ halts and rejects ω on *all* guesses $\gamma_r \in S_r$. The language L recognized by a nondeterministic formal network in time $O(R)$ is

$$L = \{\omega \mid \text{there exists an accepting guess } \gamma_r, \ r \in O(R(|\omega|))\}.$$

The function R is the amount of time required to compute the response to a given input ω, and is also the length of the guess, according to our definition.

Theorems 2 and 3 can be restated for $\mathcal{N}$ which is a nondeterministic neural network and $\mathcal{M}$ is a nondeterministic Turing machine. The proofs are similar, and thus are not provided here.

Chapter 4

Networks with Real Weights

This chapter describes the network in its full richness: the internal weights may assume any real number. One may argue that such networks are, for all practical purposes, useless since systems with infinitely precise constants cannot be built. However, the real weights are appealing for the *mathematical* modeling of analog computation that occurs in nature, as discussed in Chapter 2. In nature, the fact that the constants are not known to us, or cannot even be measured, is irrelevant for the true evolution of the system. For example, the planets revolve according to the exact values of G, π, and their masses, regardless of our inability to gauge these values. Although one could replace these constants with rational numbers in any finite time interval, and observe the same qualitative behavior, the long-term infinite-time characteristics of the system depend on the precise real values.

We prove that, although real weight neural networks are defined with unbounded precision, they demonstrate the feature, referred to as "linear precision suffices." That is, for up to q steps of computation, only the first $O(q)$ bits of both the weights and the activation values of the neurons influence the result; the less significant bits have no effect on the outcome. This means that for time bounded computation, only bounded precision is required. This property is used in chapter 10 to properly formulate the time-dependent resistance ("weak resistance") of the networks to noise and implementation error. It is interesting to note that the amount of information necessary for the neural networks is identical to the precision required by chaotic systems, therefore neural networks may constitute a framework for the modeling of physical dynamics.

The main contribution of this chapter is to show that neural networks compute functions that are not computable by Turing machines. In particular, if exponential computation time is allowed, one can specify a network for each binary language, including non-computable ones. In polynomial time, the networks compute only a very small fraction of all languages. They recognize exactly the languages belonging to the nonuniform computational class P/poly. This result can be proved by showing the polynomial equivalence either between networks and advice Turing machines (Section 1.8) or between networks and families of Boolean circuits (Boolean circuits were reviewed in Chapter 1, Section 1.8.1). It is more straightforward to show the equivalence with advice Turing machines using the machinery of the previous chapter, and we next sketch a proof along this line.

For the first side of the equivalence, let $\mathcal{M}$ be an advice Turing machine, and let $\nu(n)$ be the advice for inputs of length n; $|\nu(n)|$ is polynomial. We denote by r the concatenation

$$r = \nu(1)\nu(2)\nu(3)\cdots,$$

and by $\delta(r)$ a Cantor encoding of r.

We fix $\delta(r)$ as a weight of a neural network. The network $\mathcal{N}$ that simulates $\mathcal{M}$ can be described as a composition of two sub-networks: The first, for which $\delta(r)$ is a weight, receives the input ω. This network first stores ω and measures its length, and then retrieves the associated advice string $\nu(|\omega|)$ from $\delta(r)$. This subnetwork outputs both ω and $\nu(|\omega|)$ to the next sub-network, which performs a Turing machine simulation. The composite construction $\mathcal{N}$ simulates an advice Turing machine.

For the converse side, let $\mathcal{N}$ be a neural network with real weights. We next argue that it is polynomially included in an advice Turing machine. One might think that because advice Turing machines are very strong, they should include at least our simple network. However, note the difficulty arising from constraining the inclusion to be polynomial: given an input ω to a network $\mathcal{N}$, if $\mathcal{N}$ processes it in polynomial time, we have to compute the same using an advice Turing machine. $\mathcal{N}$ may include weights that are real numbers, specifiable by infinitely many bits; thus the activation values of the neurons may take on real values as well. How can we express real numbers of infinite precision with a Turing machine of polynomially long advice and in polynomial computation time only? For this, we use the observation "linear precision suffices," which guarantees that if a network $\mathcal{N}$ computes in time $T(n)$, then its "$T(n)$-truncated version" (to be formalized later) computes the same on any input of length n. T-truncated network can be specified with an advice of $O(T)$ bits only; this completes the simulation.

Due to the central importance of Boolean circuits in the theory of nonuniform computation, this chapter is devoted to establishing an explicit correspondence between Boolean circuits and neural networks. For functions $T, S : \mathbb{N} \to \mathbb{N}$, let NET$_R$ (T) be the class of all languages recognized by real weight neural networks in *time* T (that is, recognition of strings of length n requires at most $T(n)$) computation steps; and let CIRCUIT (S) be the class of languages recognized by nonuniform families of circuits of *size* S (that is, circuits for input vectors of length n have size at most $S(n)$). We prove the next theorem.

Theorem 4 *Let F be a function such that $F(n) \geq n$, then*

1. CIRCUIT $(F(n)) \subseteq \text{NET}_R (nF^2(n))$.

2. $\text{NET}_R (F(n)) \subseteq$ CIRCUIT $(F^3(n))$.

This chapter is organized as follows: Sections 4.1 and 4.2 contain the proof of Theorem 4, statements (1) and (2) respectively. In Section 4.3 we show the equivalence between networks and threshold circuits. As Boolean and threshold circuits are polynomially equivalent, this proof may not add any conceptually new ideas to those in previous sections: nonetheless, the direct connection and simulation between the two enables a finer comparison. Furthermore, the proof techniques differ in the two proofs. Section 4.4 makes use of results from the theory of (nonuniform) circuit complexity and states a number of corollaries, including those that relate to nondeterministic networks.

4.1 Simulating Circuit Families

We start by reducing circuit families to networks. The proof will introduce a universal architecture that can simulate all circuit families, each family requiring a different weight assignment.

Theorem 5 *There exists an integer N such that for every circuit family $\mathcal{C} = \{\mathcal{C}_n | n \in \mathbb{N}\}$ of size S that computes the language $\psi_{\mathcal{C}}$ there exists an N-processor formal real network $\mathcal{N}$ computing $\psi_{\mathcal{N}}$ so that $\psi_{\mathcal{N}} = \psi_{\mathcal{C}}$ and $T_{\mathcal{N}}(n) \in O(n\, S^2(n))$.*

The proof is provided in the remainder of this section.

4.1.1 The Circuit Encoding

Given a circuit c with size s, width w, and w_i gates in the ith level, we encode it as a finite sequence over the alphabet $\{0, 2, 4, 6\}$, as follows:

- The encoding of each level i starts with the letter 6. Levels are encoded successively, starting with the bottom level and ending with the top one.

- At each level, gates are encoded successively. The encoding of a gate g consists of three parts—a starting symbol, a 2-digit code for the gate type, and a code to indicate which gate feeds into it:

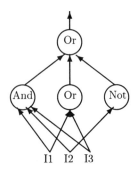

Figure 4.1: Circuit c

- It starts with the letter 0.
- A two digit sequence $\{42, 44, 22\}$ denotes the type of the gate $\{AND, OR, NOT\}$, respectively.
- If gate g is in level i, then the input to g is represented as a sequence in $\{2, 4\}^{w_{i-1}}$: the jth position in the sequence is 4 if and only if the jth gate of the $(i-1)$th level feeds into gate g.

The encoding of a gate g in level i is thus of length $(w_{i-1} + 3)$. The *length* of the encoding of a circuit c is $l(c) \equiv |\text{en}(c)| \in O(sw)$.

Example 4.1.1 The circuit c in Figure 4.1 is encoded as

$$\text{en}[c] = \mathbf{6}\underbrace{042444}_{g_1}\underbrace{044424}_{g_2}\underbrace{022242}_{g_3}\mathbf{6}\underbrace{044444}_{g_4} \ .$$

For instance, the NOT gate g_3 corresponds to the subsequence "022242": It starts with the letter 0, followed by the two digits "22" denoting that the gate is of type NOT, and ends with "242", which indicates that only the second gate of the previous level feeds into g_3.   □

We encode a nonuniform family of circuits $\mathcal{C}$, of size S, as an infinite sequence

$$e(\mathcal{C}) = 8\,\overline{\text{en}}[\mathcal{C}_1]\,8\,\overline{\text{en}}[\mathcal{C}_2]\,8\,\overline{\text{en}}[\mathcal{C}_3] \ \cdots \ , \tag{4.1}$$

where $\overline{\text{en}}[\mathcal{C}_i]$ is the encoding of $\mathcal{C}_i$ in the reversed order. This will simplify the upcoming proofs.

We can interpret formula (4.1) in base 9 and denote this representation of the family of circuits $\mathcal{C}$ as $\hat{\mathcal{C}}$,

$$\hat{\mathcal{C}} = 8\,\overline{\text{en}}[\mathcal{C}_1]\,8\,\overline{\text{en}}[\mathcal{C}_2]\,8\,\overline{\text{en}}[\mathcal{C}_3] \ \cdots \Big|_9 \ . \tag{4.2}$$

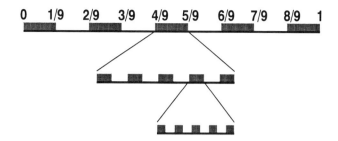

Figure 4.2: Values of the circuit encoding

The number $\hat{\mathcal{C}}$, as well as its suffixes in base 9, may assume only a restricted set of values in $[0, 1)$ as shown in Figure 4.1.1. As discussed in Subsection 3.2.1, this 9-Cantor set representation saves the need to distinguish between two very close numbers; thus the analog neurons can retrieve circuits efficiently using finite-precision operations only.

4.1.2   A Circuit Retrieval

Let $\mathcal{C}_i$ be the ith circuit in the family. We denote by $\widehat{en}[\mathcal{C}_i]$ the encoding $en[\mathcal{C}_i]$ interpreted in base 9.

Lemma 4.1.2 *For each (nonuniform) family of circuits $\{\mathcal{C}_n | n \in \mathbb{N}\}$ there exists a 16-processor network $\mathcal{N}^r$ with one input line such that, starting from the zero initial state and given the input signal*

$$u(1) = \underbrace{11 \cdots 1}_{n} 0^\infty \Big|_2 = 1 - 2^{-n}, \quad u(t) = 0 \text{ for } t > 1 \,,$$

$\mathcal{N}^r$ *outputs*

$$x_r = \underbrace{000 \cdots\cdots 0}_{2n + 2 \sum_{i=1}^{n} l(\mathcal{C}_i) + 5} \widehat{en}[\mathcal{C}_n] 0^\infty \,.$$

Proof. Let $\Sigma = \{0, 2, 4, 6, 8\}$. Denote by Δ_9 the 9-Cantor set which consists of all those real numbers q that admit an expansion of the form

$$q = \sum_{i=1}^{\infty} \frac{\alpha_i}{9^i} \tag{4.3}$$

with each $\alpha_i \subset \Sigma$. Let $\Lambda : \mathbb{R} \to [0, 1]$ be the function

$$\Lambda[x] = \begin{cases} 0 & \text{if } x < 0 \\ 9x - \lfloor 9x \rfloor & \text{if } 0 \leq x \leq 1 \\ 1 & \text{if } x > 1 \,. \end{cases} \tag{4.4}$$

Let $\Xi : \mathbb{R} \to [0,1]$ be the function

$$\Xi[x] = \begin{cases} 0 & \text{if } x < 0 \\ 2\lfloor \frac{9x}{2} \rfloor & \text{if } 0 \le x \le 1 \\ 1 & \text{if } x > 1 . \end{cases} \tag{4.5}$$

Note that for each

$$q = \sum_{i=1}^{\infty} \alpha_i/9^i \in \Delta_9 ,$$

we may think of $\Xi[q]$ as the "select left" operation, since

$$\Xi[q] = \alpha_1 ,$$

and of $\Lambda[q]$ as the "shift left" operation, since

$$\Lambda[q] = \sum_{i=1}^{\infty} \alpha_{i+1}/9^i \in \Delta_9 .$$

For each $i \ge 0$, $q \in \Delta_9$,

$$\Xi[\Lambda^i[q]] = \alpha_{i+1} .$$

The following procedure summarizes the task to be performed by the network constructed below, which in turn satisfies the requirements of the lemma.

> *Procedure* Retrieval($\hat{C}, n$)
> *Variables* counter, y, z
> *Begin*
>   counter $\leftarrow$ 0, y $\leftarrow$ 0, z $\leftarrow$ $\hat{C}$,
>   *While* counter $< n$
>   *Parbegin*
>   z $\leftarrow$ $\Lambda[z]$
>   if $\Xi[z] = 8$ then increment counter
>   *Parend,*
>   *While* $\Xi[z] < 8$
>   *Parbegin*
>   z $\leftarrow$ $\Lambda[z]$
>   y $\leftarrow$ $\frac{1}{9}(y + \Xi[z])$
>   *Parend,*
>   *Return*(y)
> *End*

The functions Λ and Ξ cannot be programmed within the neural network model due to their discontinuity. However, we can program the functions $\tilde{\Lambda}$ and $\tilde{\Xi}$, which coincide with Λ and Ξ, respectively, on Δ_9:

$$\tilde{\Lambda}[q] = \sum_{j=0}^{8} (-1)^j \sigma(9q - j) , \tag{4.6}$$

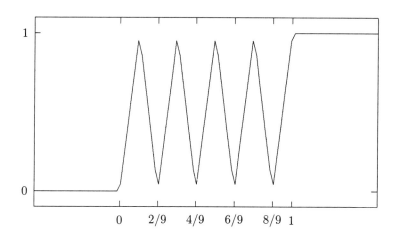

Figure 4.3: The Function $\tilde{\Lambda}[x]$.

and

$$\tilde{\Xi}[q] = 2 \sum_{j=0}^{3} \sigma(9q - (2j + 1)) . \qquad (4.7)$$

The retrieval procedure is, then, achieved by the following network:

$$
\begin{aligned}
x_i^+ &= \sigma(9x_{10} - i) \quad i = 0, \ldots, 8 \\
x_9^+ &= \sigma(2u) \\
x_{10}^+ &= \sigma(\hat{C}x_9 + x_0 - x_1 + x_2 - x_3 + x_4 - x_5 + x_6 - x_7 + x_8) \\
x_{11}^+ &= \sigma(\frac{1}{9}x_{12} + \frac{2}{9}(x_1 + x_3 + x_5 + x_7) - 2x_{13}) \\
x_{12}^+ &= \sigma(x_{11}) \\
x_{13}^+ &= \sigma(u + x_{14} + x_{15}) \\
x_{14}^+ &= \sigma(2x_{13} + x_7 - 2) \\
x_{15}^+ &= \sigma(x_{13} - x_7) \\
x_{16}^+ &= \sigma(x_{12} + x_7 - 1) .
\end{aligned}
$$

If the input u arrives at time 1, then $x_{10}(2k+3) = \tilde{\Lambda}^k[\hat{C}]$ (because of Equation (4.6)). Processors x_{13}, x_{14}, x_{15} serve to implement the counter, and processor x_{16} is the output processor. This network satisfies the requirements of the lemma. ∎

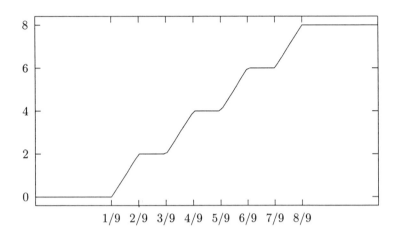

Figure 4.4: The Function $\tilde{\hat{\Xi}}[x]$.

4.1.3   Circuit Simulation By a Network

Let $\omega \in \{0,1\}^n$ be a binary sequence. Denote by en$[\omega]$ the sequence in $\{2,4\}^n$ that substitutes $(2\omega_i + 2)$ for each ω_i, and by $\widehat{en}[\omega]$ the interpretation of en$[\omega]$ in base 9, that is, en$[\omega]\big|_9$. We next construct a "universal network" for interpreting circuits.

Lemma 4.1.3 *There exists a network $\mathcal{N}^s$, such that for each circuit c and binary sequence ω, starting from the zero initial state and applying the input signal*

$$u_1 = \widehat{en}[c]\, 0\, 0 \cdots \quad u_2 = \widehat{en}[\omega]\, 0\, 0 \cdots \ ,$$

$\mathcal{N}^s$ *outputs*

$$x_0 = \underbrace{0\, 0 \cdots 0}_{T}\, y\, 0\, 0 \cdots \quad x_v = \underbrace{0\, 0 \cdots 0}_{T}\, 1\, 0\, 0 \cdots \ ,$$

where y is the response of circuit c on the input ω, and $T = O(l(c) + |\omega|)$.

Proof. It is easy to verify that, given any circuit, there is a three-tape Turing Machine that can simulate the given circuit in time $O(l(c) + |\omega|)$. This Turing machine would employ its tapes to store the circuit encoding, the input and output encoding, and the current level's calculation. Now we can simulate this machine by a rational network, as in the previous chapter. ∎

Remark 4.1.4 If the lemma only required an estimate of a polynomial number of processors, as opposed to the more precise estimate that we obtain, the proof would have followed immediately from the consideration of the *circuit value problem* (CVP). This is the problem of recognizing the set of all pairs $\langle x, y \rangle$, where $x \in \{0, 1\}^+$, and y encodes a circuit with $|x|$ input lines which outputs 1 on input x. It is known that CVP $\in$ P ([BDG90] volume I, pg 110). $\square$

4.1.4 The Combined Network

Let $\mathcal{C}$ be a circuit family. We construct a formal network and a composition of the following three networks:

- An input network, $\mathcal{N}^i$, which receives the input

$$
\begin{aligned}
u_1 &= \alpha 0^\infty \\
u_2 &= 1^{|\alpha|} 0^\infty,
\end{aligned}
$$

and computes $\widehat{en}[\alpha]$ and $u_2|_2$, for each $\alpha \in \{0, 1\}^+$. This network is trivial to implement.

- A retrieval network, $\mathcal{N}^r$, as described in Lemma 4.1.2, that receives $u_2|_2$ from $\mathcal{N}^i$, and computes $\widehat{en}[\mathcal{C}_{|\alpha|}]$. (Note that during the encoding operation, the network $\mathcal{N}^i$ produces an output of 0, and $\mathcal{N}^r$ remains in its initial state 0.)

- A simulation network, $\mathcal{N}^s$, as stated in Lemma 4.1.3, that receives $\widehat{en}[\mathcal{C}_{|\alpha|}]$ and $\widehat{en}[\alpha]$, and computes

$$
x_0 = 0^T \psi_{\mathcal{C}}(\alpha) 0^\infty \quad x_v = 0^T 10^\infty.
$$

The network $\mathcal{N}^s$ is to synchronize its inputs from the other two networks so as to read them simultaneously. The mechanism for synchronization is excluded for simplicity.

Notice that out of the above three networks, only $\mathcal{N}^r$ depends on the specific circuit family $\mathcal{C}$. Moreover, all of the weights can be taken to be rational numbers, except for the one weight that encodes the entire circuit family.

The time required to compute the response of $\mathcal{C}$ to the input ω is dominated by retrieving the circuit description. Thus, the time complexity is of the order

$$
T \in \left(\sum_{n=1}^{|\omega|} l(\mathcal{C}_n) \right).
$$

We remark that the length of the encoding $l(\mathcal{C}_n)$ is of order $O(W)S) = O(S^2)$. Since $S(n) \leq S(n+1)$ for $n = 1, 2, \dots$, we achieve the claimed bound $T = O(|\omega| S^2(|\omega|))$.

Remark 4.1.5 In case of bounded fan-in, the "standard encoding" of circuit C_n is of length $l(C_n) = O(S \log(S))$. The total running time of the algorithm is then $O(n S(n) \log(S(n)))$. □

4.2   Networks Simulation by Circuits

We next state the reverse simulation, of networks by nonuniform families of circuits.

Theorem 6 *Let $\mathcal{N}$ be a formal real network that computes in time T. There exists a nonuniform family of circuits $C = \{C_n | n \in \mathbb{N}\}$ of size $O(T^3)$, depth $O(T log(T))$, and width $O(T^2)$, that accepts the same language as $\mathcal{N}$ does.*

The proof is given in the next two subsections. In the first part, we replace a single formal network by a *family* of formal networks with small rational weights. (This is unrelated to the standard fact regarding *threshold* gates, that weights can be taken to have $n \log n$ bits.) In the second part, we simulate such a family of formal networks with circuits.

4.2.1   Linear Precision Suffices

Define a neuron to be a *designated output* processor if its activation value is used as an output of the network (i.e., it is an output processor) and it is *not* fed into any other neuron. A formal network, for which its two output neurons are designated, is called an *output designated network*. Its other neurons are considered *internal*.

For the next result, we introduce the notion of a *q-truncation* net. This is a processor (or neural) network in which the update equations take the form

$$x_i^+ = q\text{-Truncation } [\sigma(\sum_{j=1}^{N} a_{ij}x_j + \sum_{j=1}^{M} b_{ij}u_j + c_i)] \ ,$$

where q-Truncation means the operation of truncating after q bits.

Lemma 4.2.1 (Linear Precision Suffices) *Let $\mathcal{N}$ be an output designated real network. If $\mathcal{N}$ computes in time T, there exists a family of T-Truncation output designated networks $\{\mathcal{N}_n | n \in \mathbb{N}\}$ such that*

- *For each n, $\mathcal{N}_n$ has the same number of processors and input and output channels as $\mathcal{N}$ does.*

- *The weights feeding into the internal processors of $\mathcal{N}_n$ are like those of $\mathcal{N}$, but truncated after $O(T(n))$ bits.*

- For each designated output processor in $\mathcal{N}$, if this processor computes $x_i^+ = \sigma(f)$, where f is a linear function of processors and inputs, then the respective processor in $\mathcal{N}_n$ computes $\sigma(2\tilde{f} - .5)$, where $\tilde{f}$ is the same as the linear function f but applied instead to the processors of $\mathcal{N}_n$ and with weights truncated at $O(T(n))$ bits.

- The respective output processors of $\mathcal{N}$ and $\mathcal{N}_n$ have the same activation values at all times $t \leq T(n)$.

Proof. We first measure the difference (error) between the activations of the corresponding internal processors of $\mathcal{N}_n$ and $\mathcal{N}$ at time $t \leq T(n)$. This calculation is analogous to that of the chop error in floating point computation, [Atk89].

We use the following notations:

- N is the number of processors, M is the number of input lines, $L = N + M + 1$.

- W' is the largest absolute value of the weights of $\mathcal{N}$, $W = W' + 1$.

- $x_i(t)$ is the activation value of processor i of network $\mathcal{N}$ at time t.

- $\delta_w \in [0, 1]$ and $\delta_p > 0$ are the truncation errors at weights and activation values, respectively.

- $\epsilon_t > 0$ is the largest accumulated error at time t in processors of $\mathcal{N}_n$.

- $u \in \{0, 1\}^M$ is the input to both $\mathcal{N}$ and $\mathcal{N}_n$. ($u(t) = 0^M$ for $t > n$.)

- a_{ij}, b_{ij}, and c_i are the weights influencing processor i of network $\mathcal{N}$.

- $\tilde{x}_i(t)$, $\tilde{a}_{ij}$, $\tilde{b}_{ij}$, and $\tilde{c}_i$ are the respective activation values of the processors and weights of $\mathcal{N}_n$.

Network $\mathcal{N}_n$ computes at each step

$$\tilde{x}_i^+ = q\text{-Truncation } [\sigma(\sum_{j=1}^{N} \tilde{a}_{ij}\tilde{x}_j + \sum_{i=1}^{M} \tilde{b}_{ij}u_j + \tilde{c}_i)] \ .$$

We assume that up to time t, for all internal processors i, j,

$$
\begin{aligned}
|\tilde{x}_i(t) - x_i(t)| &\leq \epsilon_t \\
|\tilde{a}_{ij} - a_{ij}| &\leq \delta_w \\
|\tilde{b}_{ij} - b_{ij}| &\leq \delta_w \text{, and} \\
|\tilde{c}_i - c_i| &\leq \delta_w \ .
\end{aligned}
\tag{4.8}
$$

Using the global Lipschitz property $|\sigma(a) - \sigma(b)| \leq |a - b|$, it follows that

$$\epsilon_t \leq N(W' + \delta_w)\epsilon_{t-1} + (N + M + 1)\delta_w + \delta_p \leq LW\epsilon_{t-1} + L\delta_w + \delta_p .$$

Therefore,

$$\epsilon_t \leq \sum_{i=0}^{t-1}(LW)^i(L\delta_w + \delta_p) \leq (LW)^t(L\delta_w + \delta_p) .$$

We now analyze the behavior of the output processors. We need to prove that $\sigma(2\tilde{f} - .5) = 0, 1$ when $\sigma(f) = 0, 1$ respectively. That is,

$$f \leq 0 \Longrightarrow \tilde{f} < \frac{1}{4}$$

and

$$f \geq 1 \Longrightarrow \tilde{f} > \frac{3}{4} .$$

This happens if $|f - \tilde{f}| < \frac{1}{4}$. Arguing as earlier, the condition $\epsilon_t < \frac{1}{4}$ suffices. This is translated into the requirement

$$(L\delta_w + \delta_p) \leq \frac{1}{4}(LW)^{-t} .$$

When both δ_w and δ_p are bounded by $\frac{1}{8}(LW)^{-(t-1)}$, this inequality holds. This happens when the weights and the processor activations are truncated after $O(t \log(LW))$ bits. As L and W are constants, we conclude that a sufficient precision for a computation of length T is $O(T)$.   ∎

4.2.2   The Network Simulation by a Circuit

Lemma 4.2.2 *Let $\{\mathcal{N}_n | n \in \mathbb{N}\}$ be a family of T-Truncation output designated networks, where all networks $\mathcal{N}_n$ consist of N processors and the weights are all rational numbers with $O(T)$ bits. Then, there exists a circuit family $\{\mathcal{C}_n | n \in \mathbb{N}\}$ of size $O(T^3)$, depth $O(T \log(T))$, and width $O(T^2)$, so that $\mathcal{C}_n$ accepts the same language as $\mathcal{N}_n$ does on $\{0, 1\}^n$.*

Proof. We sketch the construction of the circuit $\mathcal{C}_n$ which corresponds to the $T(n)$-Truncation output designated net $\mathcal{N}_n$. Circuit $\mathcal{C}_n$ consists of three "subarchitectures."

1. The network $\mathcal{N}_n$ has two input lines: data and validation, where the validation line sees n consecutive 1's followed by 0's. We think of the n data bits on the data line, which appear simultaneously with the 1's in the validation line, as data input of size n. These n bits are fed

simultaneously into $\mathcal{C}_n$ via n input nodes. To simulate the sequential input in $\mathcal{N}_n$, we construct an *input-subcircuit* that preserves the input, as it is to be released one bit at a time in later stages of the computation. The input subcircuit is of size $nD_C(n)$, where D_C is the depth of the subcircuit to be described in the next item.

2. Let $p = 1, \ldots, N$ be a processor of $\mathcal{N}_n$. We associate with each processor p a subcircuit $sc(p)$. Each processor $p \in \mathcal{N}_n$ computes a truncated sum of up to $N + 2$ numbers, N of which are multiplications of two T-bit numbers. Hardwiring the weights, we can say that each processor computes a sum of $(TN + 2)$ $(2T)$-bit numbers. Using the carry-look-ahead method [Sav76], the summation can be computed by a subcircuit of depth $O(\log(TN))$, width $O(T^2N)$, and size $O(T^2N)$. (This depth is of the same order as the lower bound of similar tasks [CSV84, FSS81].) As for the saturation, one gate, p_u, is sufficient for the integer part. As only $O(T)$ bits are preserved, the activation of each processor can be represented in binary by the unit gate, p_u, and by the most significant gates

$$p_i, \; i = 1, \ldots, O(T)$$

after the operation

$$\text{AND}\,(p_i, \neg(p_u)), \; i = 1, \ldots, O(T) \; .$$

Let $sc(p')$ be a subcircuit of largest depth. Pad the other $sc(p)$'s with "demi gates" (e.g. an AND gate of a single input), so that all $sc(p)$'s are of equal depth. The output of circuit $sc(p)$ is called the *activation of* $sc(p)$.

We place the N subcircuits

$$sc(p), \;\; p = 1, \ldots, N$$

to compute in parallel. We call this subcircuit a *layer*. A layer corresponds to one step in the computation of $\mathcal{N}_n$. As $\mathcal{N}_n$ computes in time $T(n)$, $T(n)$ layers are connected sequentially. Each layer i receives the ith input bit from the input-subcircuit, and the N activation values of its preceding layer (except for layer 1, which receives input only). This *main* sub-architecture is of size $O(T^3)$, depth $O(T log(T))$, and width $O(T^2)$, where $T = T(n)$.

3. As $\mathcal{N}_n$ may compute the response to different strings of size n in different times of order $O(T)$, we construct an *output-subcircuit* that forces the response to every string of size n to appear at the top of the circuit. For each layer $i = 1, \ldots, T$, we apply the AND function to the output of the subcircuits $sc(p_1)$, $sc(p_2)$, where p_1, p_2 are the output processors

of $\mathcal{N}_n$. We transfer these values and apply the OR functions to them. The resulting value is the output of the circuit. If OR is applied at each layer, then $D_C(n)$ gates are needed for this subcircuit, where D_C is the depth of the main sub-architecture (item 2).

The resources of the total circuit are dominated by those of the main sub-architecture.   ∎

The proof of Theorem 6 follows immediately from Lemma 4.2.1 and Lemma 4.2.2.

4.3   Networks versus Threshold Circuits

A threshold circuit is defined similarly to a Boolean circuit, but the function computed by each node is now a threshold function rather than one of the Boolean functions (AND, OR, NOT). Each gate i computes

$$f_i : \{0,1\}^{n_i} \to \{0,1\} ,$$

and thus gives rise to the activation updates

$$x_i(t+1) = f_i(x_{i1}, x_{i2}, \ldots, x_{in_i}) = \mathcal{H}\left(\sum_{j=1}^{n_i} a_{ij} x_{ij}(t) + c_i\right) . \qquad (4.9)$$

Here x_{ij} are the activations of the processors feeding into it, and the a_{ij} and c_i are integer constants associated with the gate. Without loss of generality, one may assume that each of these constants can be expressed in binary form with at most $n_i \log(n_i)$ bits (see [Mur71]). If x_i is on the bottom level, it receives the external input. The function $\mathcal{H}$ is the threshold function

$$\mathcal{H}(z) = \begin{cases} 1 & z \geq 0 \\ 0 & z < 0 . \end{cases} \qquad (4.10)$$

The relationships between threshold circuits and Boolean circuits are well studied (see for example [Par94]). They are known to be polynomial equivalent in size. We provide here an alternative direct relationship between threshold circuits and real networks, without passing through Boolean circuits.

Statement of Result

Recall that NET$_R$ (T) is the class of languages recognized by formal networks (with real weights) in time T, and define T-CIRCUIT (S) as the class of languages recognized by (nonuniform) families of threshold circuits of size S. The following theorem states the polynomial equivalence of these classes.

Theorem 7 *Let F be a function such that $F(n) \geq n$, then*

1. T-CIRCUIT $(F(n)) \subseteq \text{NET}_R (nF^3(n) \log(F(n)))$.

2. $\text{NET}_R (F(n)) \subseteq T$-CIRCUIT $(F^2(n))$.

The first implication is proven similarly to the Boolean circuit case. Each threshold gate is encoded in a Cantor-like way, including the description of the weights. We next state the reverse simulation; that is, the simulation of networks by nonuniform families of threshold circuits.

Theorem 8 *Let $\mathcal{N}$ be a formal network that computes in time T. There exists a nonuniform family of threshold circuits of size $O(T^2)$, depth $O(T)$, and width $O(T)$, that accepts the same language as $\mathcal{N}$ does.*

We start with simulating $\mathcal{N}$ by the family of T-Truncation output designated networks $\{\mathcal{N}_n \mid n \in \mathbb{N}\}$ as described in Lemma 4.2.1. Next, we simulate this family of networks of depth T and size $O(T)$ via a family of threshold circuits of depth $2T$ and size $O(T^2)$.

Assume that $\{\mathcal{N}_n | n \in \mathbb{N}\}$ is the family of networks defined as the $O(T)$-truncation of $\mathcal{N}$; $\mathcal{N}_n$ has depth $T(n)$. Each neuron of $\mathcal{N}_n$ computes an addition of $N + 1$ $O(T)$-bit numbers; then it applies the σ function to it. Using a technique similar to the one provided by Parberry ([Par94], chapter 7), we show how to simulate each σ neuron of $\mathcal{N}_n$ via a threshold circuit of size $O(T)$ and depth 2. We achieve the simulation in two steps: first we add the $N + 1$ numbers and then we simulate the application of the saturation functions.

Simulating an m-truncation neuron by a threshold circuit

Step 1: Add N m-bit numbers.
Suppose the numbers are

$$z_1, \ldots, z_N ,$$

each having m-bit representation:

$$z_i = z_{i1} z_{i2} \cdots z_{im} .$$

The sum of the N m-bit numbers has $\leq m + \lfloor \log N \rfloor + 1$ bits in the representation (as the upper bound on the absolute value of the result is $N(2^m - 1)$). Generally, the sum is

$$
\begin{array}{ccccccccc}
 & & & & z_{11} & z_{12} & \cdots & z_{1m} \\
 & & & & & & \vdots & \\
+ & & & & z_{N1} & z_{N2} & \cdots & z_{Nm} \\
\hline
y_{-l} & \cdots & y_{-1} & y_0 & y_1 & y_2 & \cdots & y_m \\
\end{array}
$$

As the network is an m-truncation network, we need only to compute $y_0, y_1, \ldots y_m$. We show below how to compute y_k, $k \geq 1$. The circuit for y_0 is constructed similarly.

To compute y_k, we need to consider only z_{ij} for all i and $j \geq k$. Look at the sum:

$$
\begin{array}{ccccccc}
 & & & z_{1k} & \cdots & z_{1m} & \\
+ & & & \vdots & & & \\
\hline
 & & & z_{Nk} & \cdots & z_{Nm} & \\
\hline
c_{-l} & \cdots & c_{-1} \; c_0 & y_k & \cdots & y_m &
\end{array}
$$

It is easy to verify that

$$
\tilde{z}_k \equiv c_{-l} \cdots c_{-1} \, c_0 \, y_k \cdots y_m = \sum_{i=1}^{N} \sum_{j=k}^{m} (z_{ij} 2^{m-j}) \; .
$$

To extract from the sum the y_kth bit, we build the following circuit:

1. **Level 1:** For each possible value i of $c_{-l} \cdots c_{-1} c_0$ ($i = 1 \ldots 2^{l+1}$), we have a pair of threshold gates

$$
\tilde{y}_{ki0} = \mathcal{H}(\tilde{z}_k - c_{-l} \cdots c_{-1} c_0 \, 1 \underbrace{00 \cdots 0}_{m-k}) \; ,
$$

$$
\tilde{y}_{ki1} = \mathcal{H}(-\tilde{z}_k + c_{-l} \cdots c_{-1} c_0 \, 1 \underbrace{11 \cdots 1}_{m-k}) \; .
$$

   If $y_k = 0$, exactly one gate of each pair is active; if $y_k = 1$, only one pair has both gates active and the other pairs have only one gate active each. Thus, the y_k bit can be computed by checking if more than half of the gates in the first level are active.

2. **Level 2:** This level consists of one gate that computes the desired bit:

$$
y_k = \mathcal{H}(\sum_{i=1}^{2^{l+1}} (\tilde{y}_{ki0} + \tilde{y}_{ki1}) - (2^{l+1} + 1)) \; . \tag{4.11}
$$

Step 2: Apply the saturated function:

The value of the kth bit is

$$
b_k = \begin{cases} y_k & c_0 = 0 \\ 0 & c_0 = 1 \; . \end{cases}
$$

First, we have to compute c_0. We allocate 2^l pairs of threshold gates in the first level:

$$
\tilde{c}_{ki0} = \mathcal{H}(\tilde{z}_k - c_{-l} \cdots c_{-1} 1 \underbrace{00 \cdots 0}_{m+1-k}) \; ,
$$

$$
\tilde{c}_{ki1} = \mathcal{H}(-\tilde{z}_k + c_{-l} \cdots c_{-1} 1 \underbrace{11 \cdots 1}_{m+1-k})
$$

so that the majority function of these binary gates constitutes c_0:

$$c_0 \equiv \sum_{i=1}^{2^l} (\tilde{c}_{ki0} + \tilde{c}_{ki1}) - 2^l .$$

We change Equation (4.11) to compute b_k directly without first computing y_k:

$$b_k = \mathcal{H}(\sum_{i=1}^{2^{l+1}} (\tilde{y}_{ki0} + \tilde{y}_{ki1}) - (2^{l+1} + 1) - c_0) . \tag{4.12}$$

The size of the circuit that computes the kth bit is then $O(2^l)$, where $l = \lfloor \log N \rfloor$.

We next copy this circuit for each of the m bits. Thus, each σ neuron is simulated via a threshold circuit of depth 2 and size $O(m)$. The network itself is hence simulated via Nm copies of these circuits. Since in our case $m = O(T)$ and N is considered a constant, the simulating threshold circuit has the size $O(T^2)$ and it doubles the depth of the network $\mathcal{N}_n$.

4.4 Corollaries

Let NET_R-P and NET_R-EXP be the classes of languages accepted by formal networks in polynomial time and exponential time, respectively. Let CIRCUIT-P and CIRCUIT-EXP be the classes of languages accepted by families of circuits in polynomial and exponential size, respectively.

Corollary 4.4.1 NET_R-P $=$ CIRCUIT-P and
$\qquad\qquad\quad \text{NET}_R$-EXP $=$ CIRCUIT-EXP.

The class CIRCUIT-P is equivalent to the class P/poly and coincides with the class of languages recognized by Turing machines with advice sequences in polynomial time. The following corollary states that this class also coincides with the class of languages recognized in polynomial time by Turing machines that consult oracles (see Section 1.7), where the oracles are sparse sets. A sparse set S is a set in which for each length n, the number of words in S of length at most n is bounded by some polynomial function. For instance, any tally set, (i.e., a subset of 0^*) is an example of a sparse set. Recall from Chapter 1 that the class $P(S)$, for a given sparse set S, is the class of all languages computed in polynomial time by Turing machines that use queries from the oracle S. From [BDG90], volume I, Theorem 5.5, pg. 112, and Corollary 4.4.1, we conclude:

Corollary 4.4.2 NET_R-P $= \cup_S$ *sparse* P(S).

From [BDG90], volume I, Theorem 5.11, pg. 122 (originally, [Mul56]), we conclude as follows:

Corollary 4.4.3 NET_R-EXP *includes all possible discrete languages. Furthermore, most Boolean functions require exponential time complexity.*

The concept of a nondeterministic circuit family is usually defined by means of an extra input, whose role is that of an oracle. Similarly, a nondeterministic network was defined in Chapter 3 Section 3.8 to have an extra restricted binary input line, the *guess* line, in addition to the data and validation lines.

It is easy to see that Corollary (4.4.1), stated for the deterministic case, holds for the nondeterministic case as well. That is, if we define NET_R-NP to be the class of languages accepted by nondeterministic formal networks in polynomial time, and CIRCUIT-NP to be the class of languages accepted by nondeterministic nonuniform families of circuits of polynomial size, then:

Corollary 4.4.4 NET_R-NP$=$ CIRCUIT-NP .

Since NP $\subseteq$ NET_R-NP (one may simulate a nondeterministic Turing machine by a nondeterministic network with rational weights), the equality NET_R-NP $=$ NET_R-P implies NP $\subseteq$ CIRCUIT-P $=$ P/poly. Thus, from [KL82] we conclude:

Theorem 9 *If* NET_R-NP$=$ NET_R-P *then the polynomial hierarchy collapses to* Σ_2.

The above result suggests that a theory of computation similar to that of Turing machines is possible for our model of analog computation. Despite the very different powers of the two models, at the core of both theories is the question of whether the verification of solutions to problems is strictly faster than the process of solving them, or in other words whether P and NP are different. While our model clearly does not provide an answer, it leads us to conjecture that it is quite likely that NET_R-NP is strictly more powerful than NET_R-P .

Chapter 5

Kolmogorov Weights: Between P and P/poly

In previous chapters, we showed that the computational power of our neural networks depends on the type of numbers utilized as weights. Neural networks with rational weights, just like Turing machines, are finite objects, in the sense that they can be described with a finite amount of information. This is not true for networks with real weights; these have access to a potentially infinite source of information, which may allow them to compute nonrecursive functions. This chapter proves the intuitive notion that as the real numbers used grow richer in information, more functions become computable. To formalize this statement, we need a measure by which to quantify the information contained in real numbers.

Previous work in the field of information theory [Mar66, Mar71, Sch73] has defined the complexity of a sequence as a "measure of the extent to which a given sequence resembles a random one" [LZ76]. One particular line of research [Kol65, Cha74b, Cha74a] leads to the notion now usually referred to as the *Kolmogorov complexity* of a string. The complexity of a finite sequence α is the length of the (shortest) binary string that can be given as input to a universal algorithm in the construction (output) of α. This can be generalized in several ways to infinite strings, and hence to real numbers. Here we focus on one of the variants of the notion, called *resource-bounded* Kolmogorov complexity. This complexity is obtained by constraining not only the amount of information but also the time used by the universal algorithm to construct the numbers, thus making the notion of Kolmogorov complexity effective.

We prove that the predictability of processes such as those described by neural networks depends essentially on the resource-bounded Kolmogorov complexity of the real numbers defining the process. By gradually increasing the Kolmogorov complexity of the weights, an infinite proper hierarchy of computational classes is obtained. We find that the equivalence between neural networks and either Turing machines (for rational weights) or classes such as P/poly (for real weights) are but two special, extreme cases of the rich and detailed picture (of advice classes) spanned by the polynomial time neural networks. Among the intermediate classes we meet the Turing computable class (P/poly $\cap$ Recursive) which is strictly stronger than P. It is efficiently computed by networks having computable real weights.

This chapter builds step-by-step towards the hierarchy theorem that appears in the last section. In Section 5.1 we define Kolmogorov complexity. In Section 5.2 we focus on Turing machines that consult oracles. We prove there the main theorem of this chapter: the equivalence between neural networks of various Kolmogorov weights and Turing machines that consult corresponding oracle sets. Section 5.3 considers the relation between neural networks and Turing machines with advice. In Section 5.4 we state and prove the hierarchy theorem using the results and terminology of Section 5.2.

5.1   Kolmogorov Complexity and Reals

Kolmogorov complexity is a measure of the quantity of information contained in an individual object. It seems safe to propose that objects containing very little information would require only a brief description, and conversely more complex objects obviously require longer descriptions. A crucial point in information theory is to choose a descriptive language that will minimize redundancy. Since we are interested in descriptions that are algorithmically useful, we define a description of an object as the shortest program for a universal Turing machine that constructs the object.

Here we provide only the definitions that apply directly to our work. The historical development of Kolmogorov complexity theory, the different variants of the concept, and its many applications in several fields of Computer Science are explained in detail by Li and Vitányi [LV90, LV93]. References or proofs for the claims in the following paragraphs can be found there.

First, we review the terminology used throughout this chapter. Let α be a finite or infinite string of symbols; we denote by $\alpha_{1:k}$ the word consisting of the first k symbols of α, and by α_k the kth symbol of α. Also, $\log n$ means the function $\max(1, \lceil \log_2 n \rceil)$. Given a language L, $\chi_L \in \{0,1\}^\infty$ is its characteristic sequence, defined in the standard way: the ith bit of the sequence is "1" if and only if the ith word of Σ^* is in L (relative to some lexicographic order). Σ is taken as the smallest alphabet containing all the symbols occurring in words of L. For example, for a tally set T (see section 1.2.1), χ_T is the characteristic sequence of T relative to $\{0\}^*$.

To define the Kolmogorov complexity of real numbers, we first introduce its definition in the case of infinite binary sequences. Our definition

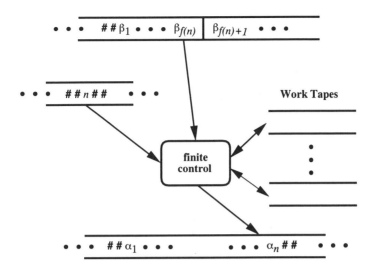

Figure 5.1: Our Universal Turing Machine

of this concept is a time-bounded analog of Kobayashi's "compressibility" version [Kob81].

Definition 5.1.1 Let U be a universal Turing machine, f and g any two functions $f, g : \mathbb{N} \to \mathbb{N}$, and let $\alpha \in \{0, 1\}^{\infty}$. We say that α has a *Kolmogorov Complexity* of $K[f(n), g(n)]$ if there exists $\beta \in \{0, 1\}^{\infty}$ such that for all but finitely many n, the universal machine U outputs $\alpha_{1:n}$ in time $g(n)$, when given $\beta_{1:f(n)}$ and n as inputs. If no condition is imposed on the running time, we say that $\alpha \in K[f(n)]$.

Figure 5.1 illustrates the definition.

Observe that here the length of the output is provided to the universal machine without being charged towards the complexity; so our definition corresponds to a complexity measure "relative to the length." The reason is that we want simple numbers (e.g., rationals) to be of extremely low complexity (e.g. constant), and allow the information contained in the length of a string to be higher. However, the definitions are equivalent (modulo small constant factors) for complexities at least logarithmic.

Generally, $K[\mathcal{F}, \mathcal{G}]$ is the set of all infinite binary sequences taken from $K[f, g]$ where $f \in \mathcal{F}$ and $g \in \mathcal{G}$. For example, a sequence is in $K[\log, P]$ if its prefixes are computable from logarithmically long prefixes of some other sequence in polynomial time.

It is easy to see that every sequence is in $K[n + O(1), \text{poly}]$; any sequence can be produced from itself, plus a constant-size program for the universal

Turing machine that copies its input to its output. A straightforward counting argument shows that there are sequences whose complexity is $K[n-O(1)]$ (i.e., they cannot be compressed beyond a constant number of bits). Such sequences are called *(Kolmogorov) random*, and the name is further justified by the fact that they pass all computable statistical tests for randomness. Informally, this means that they have all properties that can be verified algorithmically and hold with probability 1 for an infinite sequence of bits generated by tossing a fair coin. In fact, if an infinite sequence is chosen at random by such a coin-tossing process, the probability of obtaining a Kolmogorov random sequence is 1.

On the other end of the scale, sequences in $K[O(1)]$ are those that can be computed from a constant amount of information. Not surprisingly, they coincide with recursive sequences (i.e., characteristic sequences of Turing computable classes). This is easy to see in our definition, since it follows Kobayashi's variant, but is also true (although harder to show) for the standard definition [Lov69]. It is also known that all characteristic sequences of recursively enumerable sets are in $K[\log]$ [Kob81, Lov69]. So, the characteristic sequence of the halting problem is not in $K[O(1)]$, but is probably far from random.

Kolmogorov complexity was defined for symbol sequences. Now we apply this concept to real numbers by relating them in two ways to infinite binary strings. Recall our notation $\{0,1\}^\# = \{0,1\}^* \bigcup \{0,1\}^\infty$ (from Subsection 1.2.1), the two encoding functions $\delta_2, \delta_4 : \{0,1\}^\# \to [0,1]$ of Equations (3.1)and (3.2), and the range of δ_4 when restricted to the domain of infinite strings:

$$\Delta_4 = \left\{ \sum_{i=1}^{\infty} \frac{\beta_i}{4^i} \ \Big| \ \beta \in \{1,3\}^\infty \right\} .$$

The function δ_2 can be used to define the Kolmogorov complexity of numbers in $[0,1]$, and either δ_2 or δ_4 can be used to define the Kolmogorov complexity of numbers in Δ_4: a number $w \in \Delta_4$ is said to be in $K[f(n),g(n)]$ if and only if $\delta^{-1}(w) \in K[f(n),g(n)]$, where δ is either δ_2 or δ_4. In the cases where δ_2 is not injective, either of the two inverse images can be selected, since they have essentially (up to a small additive term) the same Kolmogorov complexity. The values obtained on Δ_4 by using either δ_2 or δ_4 differ only in a small constant factor, and all of our results will be independent of the function chosen.

The definition of Kolmogorov complexity is extended from the interval $[0,1]$ to $\mathbb{R}$ as follows. We say that $r \in \mathbb{R}$ is in $K[f(n),g(n)]$ if the fractional part of r is there as well; that is, we disregard the (constant) complexity of the integer part of r. Classes of real numbers $K[\mathcal{F},\mathcal{G}]$ are defined accordingly.

Finally, we use the notation $A + B$ to denote the set $\{ a + b \mid a \in A$ and $b \in B \}$, for two sets $A, B \subseteq \mathbb{R}$.

5.2 Tally Oracles and Neural Networks

The following definition will enable us to combine a fixed number of real weights into a single infinite string.

Definition 5.2.1 Let $S \subseteq \{0,1\}^{\#}$. The set S is said to be *closed under mixing* if for any finite number $k \in \mathbb{N}$ and for any k strings from S,

$$\alpha^1 = \alpha_1^1 \alpha_2^1 \alpha_3^1 \ldots, \quad \alpha^2 = \alpha_1^2 \alpha_2^2 \alpha_3^2 \ldots, \quad \ldots, \quad \alpha^k = \alpha_1^k \alpha_2^k \alpha_3^k \ldots,$$

the shuffled string

$$\alpha_1^1 \alpha_1^2 \alpha_1^3 \ldots \alpha_1^k \, \alpha_2^1 \alpha_2^2 \alpha_2^3 \ldots \alpha_2^k \, \alpha_3^1 \alpha_3^2 \ldots$$

is again an element of S.

Using the characteristic function, we can associate a real number with each tally set: the ith digit of the binary representation of the number determines if the word 0^i is in the set. Actually we have available two forms of associations: $\delta_2(\chi_T)$ and $\delta_4(\chi_T)$, the second having the property that the real number obtained belongs to Δ_4.

The following theorem states that the power of networks with real weights coincides with that of oracle Turing machines with certain tally oracles related to those weights.

Theorem 10 *Let $S \subseteq \{0,1\}^{\infty}$ be closed under mixing, and assume that $1^{\infty} \in S$. Let $\mathcal{T}$ be the following family of tally sets*

$$\mathcal{T} = \{\, T \mid \chi_T \in S \,\}.$$

Computation time in the following models is polynomially related:

1. *Neural networks whose weights are in the set $\delta_2(S) + \mathbb{Q}$.*

2. *Oracle Turing machines that consult oracles in $\mathcal{T}$.*

3. *Neural networks whose weights are in the set $\delta_4(S) + \mathbb{Q}$.*

Proof. We prove that a network in **1** can be simulated by a machine from **2**, which in turn is simulated by a network from **3**, and that the networks in **3** are a subset of those in **1**.

1 by 2: Let $\mathcal{N}$ be a network with weights in $\delta_2(S) + \mathbb{Q}$. The network has k weights, each of which can be written as the sum of a rational and a real in $\delta_2(S)$. Let these reals have base-two expansions

$$\alpha^1 = 0.\alpha_1^1 \alpha_2^1 \alpha_3^1 \ldots \;, \quad \alpha^2 = 0.\alpha_1^2 \alpha_2^2 \alpha_3^2 \ldots \;, \quad \ldots \quad \alpha^k = 0.\alpha_1^k \alpha_2^k \alpha_3^k \ldots \; .$$

Note that some of the α^i's may have two different base-two expansions (namely, ending with infinitely many "0"'s and infinitely many "1"'s). We pick one that is in S, which we know to exist because the α^i's are in $\delta_2(S)$. As S is closed under mixing, the string

$$\alpha = \alpha_1^1 \alpha_1^2 \alpha_1^3 \cdots \alpha_1^k \, \alpha_2^1 \alpha_2^2 \alpha_2^3 \cdots \alpha_2^k \, \alpha_3^1 \alpha_3^2 \alpha_3^3 \cdots$$

is again an element of S.

We show the existence of an oracle Turing machine $\mathcal{M}$ that consults a tally set with characteristic string $\chi_T = \alpha$ and simulates the network $\mathcal{N}$ with polynomial slowdown. Let R be the running time of $\mathcal{N}$, and let c be the constant provided by Lemma 4.2.1 such that precision $c \cdot R(n)$ in the weights guarantees correct results in the computation (c depends on the weights of the network). On input ω, the oracle Turing Machine $\mathcal{M}$ computes as follows:

For $t = 1, 2, 3, \ldots$ **do**

(a) **For** ($i = 1$ to $k \cdot c \cdot t$) query 0^i to T (to check the ith bit of α).

(b) Using the weights given by $\alpha_{1:kct}$, simulate $\mathcal{N}(x)$ step by step until time t.

(c) **If** ($\mathcal{N}$ has produced an output by this time), **then** output the same result and halt; **else** continue with the next t.

To see that the output is correct, note that after step (a), $\mathcal{M}$ has the weights of $\mathcal{N}$ with enough precision to correctly simulate t steps of the computation. The rational parts of the weights, being a finite, fixed amount of information, can be encoded inside the machine $\mathcal{M}$.

To conclude, note that the time overhead is polynomial in $R(|x|)$. Indeed, the simulation in step (b) takes time polynomial in t, and the time required to simulate for all $t = 1, 2, \ldots, R(|x|)$ is only quadratic in the time to simulate for $t = R(|x|)$. This quadratic overhead can be reduced to linear by trying only t's that are powers of 2.

2 by 3: This proof is based on a change in the model of oracle Turing machines. For an infinite sequence α, an α-TM is a multi-tape Turing machine equipped with a read-only input tape, a finite number of read-write tapes, and a semi-infinite read-only "reference tape" that permanently contains the infinite word α. The initial configuration has an input on the input tape, α on the reference tape, and the other tapes are blank. We sketch the proof of the following:

Lemma 5.2.2 *For any time bound $R(n) \geq n$, tally oracle Turing Machines and χ_T Turing machines are polynomially equivalent.*

Proof. Recall that χ_T is defined relative to the single letter alphabet. Essentially, the oracle Turing machine is simulated by the χ_T-TM as follows: non-oracle steps are mimicked; when a query is to be made, the machine scans the oracle tape, checking that only 0's appear there, and simultaneously advances the reference tape head. After the scan, the reference tape provides the oracle answer. If the query had an occurrence of a nonzero symbol, then the answer must be NO. The reference head is reset, and the simulation continues. For the converse case, the simulation of each move of the χ_T-TM involves a query. The oracle Turing machine calculates, on a separate tape, the correct position of the reference tape head. This is used to construct a tally query, whose answer is the bit read by the reference head. Both time overheads are easily seen to be polynomial. ∎

Now, the appropriate combination of the simulations presented in previous chapters allows us to argue the following inclusion. Fix an oracle Turing machine with oracle set $T \in \mathcal{T}$, let $\mathcal{M}$ be the corresponding χ_T-TM from Lemma 5.2.2, and let ω be an arbitrary input to $\mathcal{M}$. The simulating network acts as follows: first, a small network reads the input ω and stores its δ_4 code as the activation value of a neuron, as in Chapter 3. When the end of the input is reached, it loads $\delta_4(\chi_T)$ as the state of another neuron, through a connection with exactly that weight. Finally, it triggers another subnet that, starting with $\delta_4(\omega)$ and $\delta_4(\chi_T)$ in two specific neurons, simulates $\mathcal{M}$ on input ω and oracle T. By the construction of Chapter 3, this network exists and the simulation is efficient. Note that the only non-rational weight is $\delta_4(\chi_T) \in \delta_4(S)$.

3 by 1: No simulation is needed here. We only show that $\delta_4(S) \subseteq \delta_2(S)$, and then the stated relationship holds a fortiori. Consider the real $\delta_4(\alpha)$ for arbitrary $\alpha \in S$. Its base-four expansion uses only the digits 1 and 3, and therefore its binary expansion consists of a concatenation of pairs 01 or 11. Moreover, a pair 01 appears where α has a 0, and a pair 11 appears where α has a 1. Hence this binary expansion, seen as an infinite sequence α', is exactly the mix of $\alpha \in S$ and 1^∞. Since, by hypothesis, 1^∞ is in S and S is closed under mixing, α' is also in S. Thus $\delta_4(\alpha) = \delta_2(\alpha')$, and $\delta_4(S) \subseteq \delta_2(S)$ follows. ∎

We close this section with the following interesting observation. In many situations, we can only hope to implement networks with weights that can be effectively computed (or, communicated) using a constant amount of initial information. These are the real numbers in $K[O(1)]$ and, as explained in Section 5.1, they coincide with the recursive reals, those whose binary expansion is the characteristic sequence of a recursive set. we can now prove the following statement, which also serves as an introduction to next section.

Proposition 5.2.3 *In polynomial time, neural networks with recursive weights compute exactly* $\mathrm{P/poly} \cap \mathrm{REC}$, *the recursive part of* $\mathrm{P/poly}$.

The class $\mathrm{P/poly} \cap \mathrm{REC}$ is different than P, as it strictly includes P, and hence this corollary says that the networks demonstrate speedup without the use of any nonrecursive weights.

Proof. Let S be the set of recursive infinite sequences. It is easy to see that it is closed under mixing, and of course it contains 1^{∞}. Then $\delta_2(S) + \mathbb{Q}$ is exactly the set of recursive reals. By theorem 10, neural networks with recursive weights are polynomially equivalent to oracle Turing machines which consult recursive tally sets as oracles. We next prove that such oracle Turing machines compute the class $\mathrm{P/poly} \cap \mathrm{REC}$. For this proof we use the result from ([BDG90] I) that $\mathrm{P/poly}$ is the computational class of tally oracle Turing machines that computes in polynomial time.

Let T be recursive tally and $\mathrm{P}(T)$ be the class of languages decidable in polynomial time by an oracle Turing machine querying the oracle T. Let L be in $\mathrm{P}(T)$. L is recursive because T is so, it is also in $\mathrm{P/poly}$ because it is accepted by a tally oracle machine in polynomial time.

Conversely, assume a language $L \in \mathrm{REC} \cap \mathrm{P/poly}$. Because it is in $\mathrm{P/poly}$ it must be accepted by an oracle Turing machine $\mathrm{P}(T)$ for some tally set T [BDG90]. We next prove that because L is recursive, there is also a recursive tally set T' for which $L \in \mathrm{P}(T')$. Here we sketch one possible construction of T'.

Assume that $L \in \mathrm{P}(T)$ is witnessed by a Turing machine $\mathcal{M}$ running in time $p(n)$. We say that a string α of length $p(n)$ is good advice for length n if, for every ω of length n, $\mathcal{M}(\omega)$ correctly decides whether $\omega \in L$ when using any tally set whose characteristic sequence has α as the first $p(n)$ bits. Note that $p(n)$ bits are enough for the simulation, and that to decide this property on α it is enough to apply the decision procedure for L 2^n times. There exists good advice of length $p(n)$, an example is the prefix of χ_T, but we choose the lexicographically first good advice for length n, calling it α_n. The tally set T' is defined as

$$T' = \{\langle 0^n, 0^i \rangle \mid n \geq 0,\ 1 \leq i \leq p(n),\ \text{and the } i^{th} \text{ bit of } \alpha_n \text{ is } 1\}$$

where the pairing function $\langle ., . \rangle$ is taken to map $0^* \times 0^*$ to 0^*; T' has $p(n)$ tally words. A straightforward modification of $\mathcal{M}$ accepts L relative to T', and it is also easy to prove that T' is recursive. This T' is easily constructed with complexity at most exponential in L, but better bounds on its complexity can be obtained [Sch86]. ∎

5.3 Kolmogorov Weights and Advice Classes

Our purpose here is to show that the Kolmogorov complexity of the network weights relates to the structural notion of the amount of advice in nonuniform classes.

We prove this claim in a general setting, so as to immediately include several cases of interest. In order to do this, we introduce some technical conditions that we expect all of our advice bounds to have; namely, we say that a class of functions $\mathcal{F}$ forms *reasonable advice bounds* if
1) $\mathcal{F} \subseteq P$,
2) it is closed under $O(\cdot)$, and
3) for every polynomial p and every $f \in \mathcal{F}$ there exists $g \in \mathcal{F}$ such that $f \circ p \leq g$.
A number of reasonable advice bounds are described below, and all advice bounds used in the literature are included in our definition of reasonable advice.

In order to apply Theorem 10 to classes of the form $K[\mathcal{F}, P]$, we first show that they are closed under mixing. Here we use the fact that our definition of Kolmogorov complexity is relative to the input length.

Lemma 5.3.1 *Let $\mathcal{F}$ be a function class that is closed under $O(\cdot)$. Then, $K[\mathcal{F}, P]$ is closed under mixing.*

Proof. Let $k \in \mathbb{N}$ and $\alpha^1, \ldots, \alpha^k \in K[\mathcal{F}, P]$. That is, for all $i = 1, \ldots, k$, $\alpha^i_{1:n}$ is computed in polynomial time $g_i(n)$ using $\beta^i_{1:f_i(n)}$ as input. Thus, the shuffled string of the α^i's ($\tilde{\alpha}$) up to length n can be computed in time

$$g(n) = O(n) + \sum_{i=1}^{k} g_i(\lceil \frac{n}{k} \rceil)$$

from an input that is equal to the prefix of length

$$f(n) = \sum_{i=1}^{k} f_i(\lceil \frac{n}{k} \rceil)$$

of the shuffled string of the β^i's. It is easy to verify that as both $\mathcal{F}$ and P are closed under $O(\cdot)$, g is polynomial and $f \in \mathcal{F}$. Thus, $\tilde{\alpha} \in K[\mathcal{F}, P]$. ∎

Now we are ready to prove the relation between the advice classes and languages accepted by neural networks. Our notion of advice classes is based on prefix advice, as introduced in Definition 1.8.3: an advice ν has the prefix form, i.e. ν_n is the prefix of ν_m for $n < m$. In addition, ν_n is good for all input strings $|\omega| \leq n$ and not only for these with $|\omega| = n$.

Theorem 11 *Let $\mathcal{F}$ be a class of reasonable advice bounds. Then the class* $\mathrm{P}/\mathcal{F}^*$ *is exactly the class of languages accepted by polynomial time neural networks with weights in $K[\mathcal{F}, \mathrm{P}]$.*

Proof. Note that, since all constants are in $O(\mathcal{F}) = \mathcal{F}$ and every rational has constant complexity, $\mathbb{Q} \subseteq K[\mathcal{F}, \mathrm{P}]$. For the same reason, 1^∞ is in $K[\mathcal{F}, \mathrm{P}]$. Furthermore, by Lemma 5.3.1, the class $K[\mathcal{F}, \mathrm{P}]$ is closed under mixing. By Theorem 10, it suffices to show that $\mathrm{P}/\mathcal{F}^*$ coincides with the class of languages decidable by Turing machines in polynomial time with tally oracle sets T such that $\chi_T \in K[\mathcal{F}, \mathrm{P}]$.

Thus, assume that $L \in \mathrm{P}(T)$ with such a T, and consider an infinite string α such that the first m bits of χ_T can be recovered from the first $f(m)$ bits of α in polynomial time in m, with $f \in \mathcal{F}$. Let p be the polynomial bounding the time necessary to decide L, and let $g \in \mathcal{F}$ be a bound on $f \circ p$ such that $g(n)$ is computable in polynomial time in n. This g exists by the assumption that $\mathcal{F}$ forms reasonable advice bounds. Then L is in $\mathrm{P}/\mathcal{F}^*$ by choosing as advice for length n the first $g(n)$ bits of α, from which up to $p(n)$ bits of χ_T can be reconstructed in polynomial time and then used to decide L instead of querying the oracle set T.

Conversely, let $L \in \mathrm{P}/\mathcal{F}^*$ via the polynomial time machine $\mathcal{M}$ and let α be the infinite word whose prefix of length $f(m)$ can be used as an advice string for length m, with $f \in \mathcal{F}$. Since $\mathcal{F}$ is reasonable, we can take $g \in \mathcal{F}$, $g > f$ that is easily computable from m, and is bounded by a polynomial p.

Let $q(t) = (t+1)p(t)$, so that for all t greater than some n_0, $p(t+1) \leq q(t+1) - q(t)$. In this case

$$(g(t+1) - g(t)) \leq g(t+1) \leq p(t+1) \leq (q(t+1) - q(t)).$$

Consider the oracle tally set T whose characteristic function is defined as follows: bits from 1 to $g(1)$ of χ_T are the first $g(1)$ bits of α; bits $g(1) + 1$ to $q(1)$ of χ_T are zero; when $1 \leq \ell \leq (g(t+1) - g(t))$, bit $q(t) + \ell$ of χ_T is bit $g(t) + \ell$ of α; and when $(g(t+1) - g(t)) < \ell \leq (q(t+1) - q(t))$, bit $q(t) + \ell$ of χ_T is 0. It is easy to see that, given m and $g(m)$ bits of α, we can print out $q(m) \geq m$ bits of χ_T in polynomial time. So, $\chi_T \in K[\mathcal{F}, \mathrm{P}]$, and in polynomial time a Turing machine with oracle T can obtain the necessary advice words to simulate $\mathcal{M}$ and thus decide L. This completes the proof. ∎

Some interesting special cases arise when considering various natural bounds for the Kolmogorov complexity:

- $S = K[\mathrm{poly}, \mathrm{P}]$, the set of arbitrary strings. The class of languages accepted in this case is P/poly; this is the main result of Chapter 4.

- $S = K[\log, P]$. In this case, the class of languages accepted is $P/\log^*$, which equals strong-$P/\log$ (see section 1.8 just above Def 1.8.3) as shown in [BDG90, Her96].

- $S = K[O(1), P]$, that is, characteristic strings of recursive sets computable in polynomial time. Here it is important that we use Kobayashi's definition, since with other variants the class $K[O(1), P]$ may contain characteristic functions of recursive sets requiring an arbitrarily long time to decide. The class of languages accepted in this case is P, since by the results of Chapter 3, polynomial time computations are exactly those of polynomial-time neural networks with rational weights.

5.4   The Hierarchy Theorem

In this section we show the existence of a proper hierarchy of complexity classes of networks. We first define an ordering between classes of tally sets; then we show that oracle Turing machines consulting these classes of oracles result in an infinite hierarchy of computational classes. Finally, we re-apply the characterization in Theorem 10 and obtain the desired hierarchy of neural networks. The advantage of this approach is that Kolmogorov complexity allows for a simple argument to prove that all classes defined by oracle Turing machines are indeed different.

For the definition of *partial order* between function classes we need the concept $o(\cdot)$ of order of magnitude, which is different than the $O(\cdot)$ described in Chapter 1. $o(f)$ is the set of functions g such that, for every $c > 0$ and for all but finitely many n, $g(n) < cf(n)$. The set $o(\cdot)$ coincides with the set of functions g such that

$$\lim_{n \to \infty} \frac{g(n)}{f(n)} = 0 \ .$$

Our (strict) partial order on function classes is defined as follows. Let $\mathcal{F}$ and $\mathcal{G}$ be function classes. We say that $\mathcal{F} \prec \mathcal{G}$ if there is some nondecreasing function $s(n) \in \mathcal{G}$, computable in time polynomial in n, so that $s(n) = o(n)$, and for every polynomial p and every $r \in \mathcal{F}$, $r \circ p = o(s)$. (Note, the computation of $s(n)$ is a kind of subroutine in a larger program that works on strings of length n and $s(n)$, not $\log(n)$; also note that $\mathcal{F}$ is sublinear by transitivity.)

This partial order defines an infinite hierarchy. For instance, one may consider the function classes $\theta_i = \{q_1, \ldots, q_i\}$, where $q_i = \log^{(i)}$ is defined inductively by $q_1 = \log$ and $q_i = \log(q_{i-1})$, for $i > 1$.

For the next theorem, we denote by $\mathcal{T}_{\mathcal{F}}$ the family of all tally sets T with the property that $\chi_T \in K[\mathcal{F}, P]$, and by $P(\mathcal{T}_{\mathcal{F}})$ the class of all sets decidable in polynomial time by an oracle Turing machine with an oracle in $\mathcal{T}_{\mathcal{F}}$.

Theorem 12 *Let $\mathcal{F}$ and $\mathcal{G}$ be non-empty function classes, such that $\mathcal{F} \prec \mathcal{G}$. Then,* $\mathrm{P}(\mathcal{T}_\mathcal{F})$ *is properly included in* $\mathrm{P}(\mathcal{T}_\mathcal{G})$.

Proof. Let $s(n) \in \mathcal{G}$ be a bound on $\mathcal{F}$ as in the definition of partial order. Note that, since $\mathcal{F}$ is nonempty, s must be unbounded. Trivially, $\mathrm{P}(\mathcal{T}_\mathcal{F}) \subseteq \mathrm{P}(\mathcal{T}_{\{s\}}) \subseteq \mathrm{P}(\mathcal{T}_\mathcal{G})$. We define a set L that is in $\mathrm{P}(\mathcal{T}_\mathcal{G})$ but not in $\mathrm{P}(\mathcal{T}_\mathcal{F})$. Choose an infinite sequence $\gamma \notin K[\frac{n}{2}]$. (Such sequences exist as discussed in Section 5.1.) For each n define the string β_n as

$$\beta_n = \gamma_{1:\frac{s(n)}{2}} 0^{n - \frac{s(n)}{2}} \ ,$$

if $n \geq \frac{s(n)}{2}$, and $\beta_n = 0^n$ otherwise. Let L be the tally set with characteristic string $\beta_1 \beta_2 \beta_3 \ldots$. Given $\gamma_{1:\frac{s(n)}{2}}$ and recalling that $s \in \mathcal{G}$, it is easy to build $\chi_{A_{1:n}}$, so that

$$\chi_L \in K[\frac{s}{2} + c, \mathrm{P}] \subseteq K[s, \mathrm{P}] \ .$$

However, $L \notin \mathrm{P}(\mathcal{T}_\mathcal{F})$. If we assume otherwise, then there is some machine that accepts L in time p_1 with a tally set T as oracle, where $\chi_T \in K[r, p_2]$, p_1 and p_2 are polynomials, and $r \in \mathcal{F}$. In time $p_1(n)$, the machine can query at most the first $p_1(n)$ elements of T. Using this machine, we can easily print out (in time $np_1(n)$) $\beta_1 \beta_2 \ldots \beta_n$; and hence we can also print $\gamma_{1:\frac{s(n)}{2}}$, given inputs $\frac{s(n)}{2}$ and the first $r(p_1(n)) + O(1) < \frac{s(n)}{4}$ bits of the seed for χ_T. This contradicts the choice of γ.   ∎

Remark 5.4.1 The requirement that s be nondecreasing can be removed at the expense of some technical complications. For instance, the upper bound on the complexity χ_L would only hold infinitely often, and γ can be chosen to be highly complex almost everywhere.

Let us note that there is nothing special about the function class P in the previous theorem; similar separation results can be proved for other well-behaved families of run-time functions.

As we have seen, Theorem 10 establishes a connection between the set of weights of a family of networks and the oracle Turing machines that consult related tally sets. Theorem 12 displays a hierarchy of oracle Turing machines querying oracles that belong to different Kolmogorov complexity classes. From both theorems, we immediately conclude the following:

Theorem 13 (Hierarchy Theorem) *Let $\mathcal{F}$ and $\mathcal{G}$ be two function classes closed under $O(\cdot)$, with $\mathcal{F} \prec \mathcal{G}$. Let* NET $_{K[\mathcal{F},\,\mathrm{P}\,]}$ *be the class of languages accepted by networks that compute in polynomial time, and each of which uses weights from $K[\mathcal{F},\mathrm{P}]$, and similarly for $\mathcal{G}$. Then:*

$$\mathrm{NET}\ _{K[\mathcal{F},\,\mathrm{P}\,]} \subsetneq \mathrm{NET}\ _{K[\mathcal{G},\,\mathrm{P}\,]} .$$

Chapter 6

Space and Precision

So far, we have considered neural networks with two types of resource constraints: time, and the Kolmogorov complexity of the weights. Here, we consider rational-weight neural networks in which a bound is set on the precision available for the neurons. The issue of precision comes up when simulating a neural network on a digital computer. Any implementation of real arithmetic in hardware will handle "reals" of limited precision, seldom larger than 64 bits. When more precision is necessary, one must resort to a software implementation of real arithmetic (sometimes provided by the compiler), and even in this case a physical limitation on the length of the mantissa of each state of a neural network under simulation is imposed by the amount of available memory. This observation suggests that some connection can be established between the space requirements needed to solve a problem and the precision required by the activations of the neural networks that solve it.

In this chapter we show the analogy between the space S required by a Turing machine and the total precision p required by the neurons of an equivalent network. This analogy is demonstrated by two neural network models. In Section 6.1 we prove that if the number of neurons is constant, then $p(n) = O(S(n))$. In Section 6.2 we consider neurons of constant precision and allow for a growing network. *(This is the only place in this book where we consider variable sized networks.)* We prove that the size of such a network that simulates a Turing machine of space S is $O(S)$ as well.

6.1 Equivalence of Space and Precision

Definition 6.1.1 A rational neural network *works within precision $S(n)$* if and only if all of the weights and all of the rational activation values of the neurons, throughout a computation on an input of length n, can be represented in binary form with $O(S(n))$ bits.

We observe the following:

Theorem 14 *Let $S(n) \geq n$ be a space-constructible function. Then the following statements are equivalent:*

1. *The language L is accepted by a Turing machine using space $O(S(n))$.*

2. *The language L is accepted by a neural network working within precision $O(S(n))$.*

The proof is omitted as it is similar to the proof of Theorem 2 in Chapter 3. We point out that the proof relies on a preliminary phase in which the input is completely loaded into the state of a specific neuron, before proceeding to the actual computation. To store the input, this neuron requires linear precision, which is the reason for the condition $S(n) \geq n$.

It is also interesting to see what happens under *sublinear* precision bounds. For this purpose we need to abandon the "network without inputs" described in Chapter 3 Theorem 3, and return to the original convention of inputs through an input stream introduced in Chapter 2. Similarly, in the classical theory when considering the space resource, a particular type of Turing machine is used. These have a designated input tape which is read-only, in addition to working tapes. The space used by the machine is the number of work tape cells required for the computation. We distinguish between three types of such Turing machines. When no restriction is placed on the movement of the input head, the machine is called an *off-line* Turing machine. In an *on-line* Turing machine, the input head can either move to the right or stay still, but cannot backtrack to the left. We define here *lr* Turing machines as the subclass of on-line Turing machines that must move their input head to the right one symbol per step. In both on-line models, it is allowed to continue working after exhausting the input. This last stage of work uses only the information gathered in the work tapes during the reading. Clearly these restricted model are equivalent to the standard model for space bounds which are at least linear.

If the precision of the neuron which holds the input is ignored, then theorem 14 would hold for sublinear space bounds as well. We prefer to consider the precision required for all of the neurons. For sublinear $S(n)$, the next result provides one necessary and one sufficient characterization of the connection between precision bounds in neural networks and space constraints in Turing machines.

Theorem 15 *Let $S(n)$ be any space-constructible function.*

1. *If a language L is accepted by an lr-machine using space $O(S(n))$, then L is accepted by a neural network working within precision $O(S(n))$.*

2. If a language L is accepted by a neural network working within precision $O(S(n))$, then L is accepted by an on-line machine using space $O(S(n))$.

The first part of the theorem essentially corresponds to proving that for the simulation of an lr-machine, the intermediate step of loading the input into a single neuron state (as done in Section 3.6) is not necessary. The second part of the theorem consists of a straightforward simulation of the computation of the neural network. The state of each of the fixed number of neurons is kept on a worktape, where it is guaranteed to fit due to the precision bound. Since the network receives its input in real-time, there is never a need to backtrack the input head during the simulation. Observe, however, that the simulating machine is not an lr-machine, since each update of the network may require a non-constant number of Turing machine steps.

Off-line space-bounded machines can be proven equivalent to precision-bounded neural networks under the following input convention.

Definition 6.1.2 A neural network with *cyclic input* receives the input w through two input lines as follows: the data line brings in the bits of the input w repeatedly, w^∞, and the validation line brings in $(`1"`0`^{|w|-1})^\infty$.

In this definition, the data line carries the stream $wwwwww\cdots$, and the validation line, instead of marking the end of the input, marks the beginning of each cycle. With this (admittedly somewhat artificial) input convention we obtain the following theorem.

Theorem 16 Let $S(n) \geq \log n$ be a space-constructible function. Then the following are equivalent.

1. The language L is accepted by an off-line Turing machine using space $O(S(n))$.

2. The language L is accepted by a neural net with cyclic input working within precision $O(S(n))$.

Here we only sketch the proof.

Proof. $1 \Rightarrow 2$)
The network $\mathcal{N}$ simulating the Turing machine $\mathcal{M}$ is built conceptually of two subnetworks. One subnetwork of constant size receives as input the bit currently being scanned by the input-tape head of $\mathcal{M}$ and the state of $\mathcal{M}$, and returns a new state and the direction in which to move the input-tape head. One of the neurons keeps a rational value that, interpreted as an integer value, indicates the current position of the input-tape head; its value is incremented or decremented, depending on the direction of movement. The

other subnetwork, triggered by the value "1" that marks the beginning of
each cycle, counts up to the position of the input-tape head to detect the
input symbol necessary for the simulation of the next step. With additional
time for preprocessing, it is possible to do the counting in real-time using only
logarithmic precision.

$2 \Rightarrow 1$)
For the reverse implication, use the same simulation as for the on-line case.
When reaching the right end of the input, stop the simulation, reset the input
tape head, and resume it; when the simulating machine reads the first symbol
of the input, it simulates a "1" on the input validation line.   ■

Taking into account the fact that time-bounded rational networks correspond modulo polynomial-time simulations to time-bounded Turing machines
(Chapter 3), together with Theorem 14 here, allows us to close this section
by pointing out a remark on the "linear precision suffices" Lemma 4.2.1. If
we restrict this lemma to rational weights it corresponds in some sense with
the basic theorems relating time-bounded and space-bounded classes, and in
particular to the previous result, that everything done in time $T(n)$ is done in
space $T(n)$ as well. The "linear precision" lemma, restricted to the rational
case, would be essentially the neural network equivalent of this result.

6.2   Fixed Precision Variable Sized Nets

A rational network *works within a constant precision* if its precision bound is
a constant, independent of the input size.

Clearly, a finite network that works within a constant precision is essentially a finite automaton. To gain recursive power with neurons of constant
precision, we have to allow the network to grow as a function of the computation time.

The number of neurons required to simulate a Turing machine that operates in space S is at least linear in S. We prove that this lower bound
is indeed achievable. Furthermore, we sketch the construction of a network
which, using this number of neurons, simulates a Turing machine in linear
time.

Theorem 17 *Let $\mathcal{M}$ be a Turing machine that computes a function ψ using
space S and time T. Then, for each positive integer p, there exists a network
$\mathcal{N}$ with $\frac{cS}{p}$ neurons (c is a constant independent of p) that computes the same
function ψ in time $O(T)$, and works within constant precision p.*

Proof. Assume that $\mathcal{M}$ is a stack machine. We show how to simulate a binary
stack of $\mathcal{M}$ with p-precise neurons. This, combined with the proof of Theorem
2 in Chapter 3, implies the theorem.

Let $\alpha \in \{0,1\}^*$; we define the two p sequences

$$S(\alpha) = \{S_n\}_{n=1}^{\lceil \frac{|\alpha|}{p} \rceil + 1} \quad \text{and} \quad \bar{S}(\alpha) = \{\bar{S}_n\}_{n=1}^{\lceil \frac{|\alpha|}{p} \rceil + 1}$$

as follows:

$$S_n(\alpha) = \begin{cases} \alpha_{(n-1)p+1} \cdots \alpha_{np} & n < \lceil \frac{|\alpha|}{p} \rceil \\ \alpha_{(n-1)p+1} \cdots \alpha_{|\alpha|} 0^{p\lceil \frac{|\alpha|}{p} \rceil - |\alpha|} & n = \lceil \frac{|\alpha|}{p} \rceil \\ 0^p & n = \lceil \frac{|\alpha|}{p} \rceil + 1, \end{cases}$$

and $\bar{S}_n(\alpha) = \beta_1\beta_2 \cdots \beta_p$ when $S_n(\alpha) = \beta_p \cdots \beta_2\beta_1$.

Given a binary stack, we identify it with a binary sequence $\alpha = \alpha_1\alpha_2 \cdots \in \{0,1\}^*$ in top to bottom order. We then encode α by the encoding function $\delta_4(\alpha)$ (from Equation (3.2)) and associate with it the two following sequences of $\lceil \frac{|\alpha|}{p} \rceil + 1$ neurons:

$$\{g_n \mid g_n = S_n(\alpha)|_4\} \quad \text{and} \quad \{\hat{g}_n \mid g_n = \bar{S}_n(\alpha)|_4\} .$$

Three additional neurons are associated with each g_n and $\hat{g}_n$: $\zeta(g)$ and $\tau(g)$ compute the top and non-empty of the neuron g, respectively, as described in Equations (3.5) and (3.6). The third additional neuron κ computes

$$\kappa[\zeta[g], \tau[g]] = \begin{cases} 0 & \zeta[g] = 0 \\ \frac{1}{4} & \zeta[g] = 1, \tau[g] = 0 \\ \frac{3}{4} & \tau[g] = 1 . \end{cases}$$

Using these neurons, the stack operations **push** and **pop** can each be simulated in a constant number of steps.

- **Push(I)**, $I \in \{\frac{1}{4}, \frac{3}{4}\}$:

$$g_1 = \frac{1}{4}(g_1 - \frac{\kappa[\hat{g}_1]}{4^{p-1}}) + I$$

$$g_n = \frac{1}{4}(g_n - \frac{\kappa[\hat{g}_n]}{4^{p-1}}) + \kappa[\hat{g}_{n-1}] \quad n > 1$$

$$\hat{g}_1 - 4(\hat{g}_1 \quad \kappa[\hat{g}_1] + \frac{I}{4^p})$$

$$\hat{g}_n = 4(\hat{g}_n - \kappa[\hat{g}_n] + \frac{\kappa[\hat{g}_{n-1}]}{4^p}) .$$

- **Pop**:

$$g_n = 4(g_n - \kappa[g_n] + \frac{\kappa[g_{n+1}]}{4^p})$$

$$\hat{g}_n = \frac{1}{4}(\hat{g}_n - \frac{\kappa[g_n]}{4^{p-1}}) + \kappa[g_{n+1}] .$$

Note that one can reduce the number of neurons; we only provided an intuitive construction to argue the existence of a network with $O(\lceil \frac{S}{p} \rceil])$ neurons and linear slowdown.   ∎

To summarize, in this chapter we have formally shown the intuitive equivalence between the space resource of a Turing machine and the total precision used by an analog neural network.

Chapter 7

Universality of Sigmoidal Networks

Up to this point we considered only the saturated linear activation function. In this chapter, we investigate the computational power of networks with *sigmoid* activation functions, such as those widely considered in the neural network literature, e.g.,

$$\varrho(x) = \frac{1}{1 + e^{-x}}$$

or

$$\varrho(x) = \frac{2}{1 + e^{-x}} - 1. \tag{7.1}$$

In Chapter 10 we will see that a large class of activation functions, which also includes the sigmoid, yields networks whose computational power is bounded from above by P/poly. In this chapter we obtain a lower bound on the computational power of sigmoidal networks. We prove that there exists a universal architecture of sigmoidal neurons that can be used to compute any recursive function, with exponential slowdown. Our proof techniques can be applied to a much more general class of "sigmoidal-like" activation functions, suggesting that Turing universality is a common property of recurrent neural network models. In conclusion, the computational capabilities of sigmoidal networks are located in between Turing machines and advice Turing machines.

For the Turing machine simulation, we must contend with substantial technical difficulties that arise when using the activation function ϱ. The saturated linear activation function σ has two properties not shared by the sigmoid, which were useful in the Turing machine simulation. The function σ has an interval on which it computes the identity function. This property facilitates long term memory of the neural activation, and was used to store the contents of the Turing machine tape. When using the sigmoid ϱ, the memory property is lost because the iterates of ϱ, $\varrho^i(x)$ have different values for all i (for almost all x). In addition, ϱ saturates to two different *exact* values. This was used in the implementation of logical operations which are the heart of the control of a Turing machine. By simply multiplying the input to a neuron, we could map slightly "noisy Boolean inputs" (e.g., 0.2, 0.98) to exact Boolean outputs. The sigmoid activation function does not have this perfect saturation property. As a result, the neurons cannot exactly compute the logical AND function, even for inputs x and y which are guaranteed to have $\{0,1\}$ values. Not only will the answer not take on an exact $\{0,1\}$ value, but it will take on slightly different values, depending on whether $(x,y)=$ (0,0) or $(x,y)=$ (0,1). We guarantee only that these two results of the AND operations will be reasonably close to 0. Such an approach can be used to simulate finite automata: we can consider a neighborhood of 0 as the value 0 and a neighborhood of 1 as the value 1. It is not applicable, however, for implementing automata that remember state information of unbounded size (e.g. the tape of a Turing machine).

We solve the problem of memory in a sigmoidal network by introducing a new type of automaton, called an *alarm clock machine*, that does not rely on remembering state information. This machine, which is shown to be Turing universal, will be simulated by a sigmoidal neural network. An alarm clock machine consists of a restricted finite control that has access to a finite number of alarm clocks. Until it is woken by one or more alarm clocks, the finite control spends its time in a memoryless "sleep" state. When woken, the finite control is allowed to run for a *constant* number of steps. Since it remembers nothing from before it woke up, it must base its actions only on its knowledge of which clocks alarmed during this short waking phase (see Figure 7.1).

Unbounded memory (the tape of a Turing machine for example) is encoded in the temporal pattern of the alarms. Each alarm clock c_i has a variable period (also referred to as "day") p_i. If the clock alarms at time t_i, then the clock will next alarm at time $t_i + p_i$. Memory is manipulated by the controller, which may perform operations such as delaying a clock (so that it next alarms at $t_i + 1$), or lengthening the clock's day (setting $p_i = p_i + 1$), before going back to sleep.

The alarm clocks store information in the frequency domain. Unlike the Turing machine, in which strings are encoded in binary form, alarm clock machines use the numerical value associated with a string (for example, by any of the δ encoding functions from Chapter 3). We thus lose the compact (logarithmic) representation, causing an exponential slowdown in the simulation of a Turing machine.

The sigmoidal network that is used to simulate an alarm clock machine consists of three parts. One is devoted to the simulation of the alarm clocks. Each alarm clock can be viewed as a pair of coupled oscillators (these will be named restless counters) which can be implemented by sigmoidal neurons.

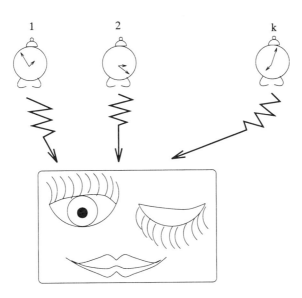

Figure 7.1: An Alarm Clock Machine: k alarm clocks and a controller that is in a sleeping state and is active for only c steps when woken up

Another part of the network simulates the finite control. The finite control neurons in this implementation spend most of their time in a low-noise sleep state that prevents the state of the clocks from being corrupted faster than it can be repaired. The third part is a coupling device that receives signals from the finite control and transfers "cleaner" signals to the alarm clocks. Each alarm clock has its own coupling device, implemented by a sigmoidal neuron. As a dynamical system, a sigmoidal neuron has two stable attracting fixed points. As long as the finite control is in its low-noise sleep state, this neuron converges to a fixed point, which has the effect of "cleaning" the signals from the finite control.

Most of this chapter is devoted to the construction of this three-component sigmoidal network simulating alarm clock machines. In the proof we use one non-rational weight. It is still unknown whether the same result can be obtained using rational weights only. In Section 7.1 we introduce alarm clock machines and prove that they are Turing universal. In Section 7.2 we show how to substitute the alarm clocks with "restless counters" that behave in a restricted manner. In Section 7.3 we describe how to simulate alarm clock machines (with restless counters) by sigmoidal first-order neural networks, and thus prove the universality of the networks. In Section 7.4 we generalize the universality result to other "sigmoidal-like" networks.

7.1   Alarm Clock Machines

An *alarm clock machine* $\mathcal{A}$ is a triplet (F, k, c) where $k, c \geq 1$ and F is a function from $\{0, 1\}^{kc}$ to a subset of ACTION, where

$$\text{ACTION} = \{delay(i), lengthen(i)|1 \leq i \leq k\} \cup \{\text{halt}\}.$$

Here k denotes the number of alarms clocks available to $\mathcal{A}$, and F is a function that, based on the history of alarms from the last c time steps, halts and/or performs some simple operations on its clocks.

The input to (F, k, c) consists of $((p_1, t_1), \ldots, (p_k, t_k))$, where p_i denotes the period of clock i, and time t_i denotes the next time it is set to alarm. For notational ease, we (conceptually) keep arrays $a_i(t)$, for $t \in \mathbb{N}$ and $1 \leq i \leq k$, with each entry initially set to "0". The alarm clock machine operates as follows. At time step T (initially 0), for $1 \leq i \leq k$, if $t_i = T$, then $a_i(T)$ is set to "1" and t_i is set to $t_i + p_i$. This event corresponds to clock i alarming. F then looks at $a_i(t)$ for $1 \leq i \leq k$ and $T - c < t \leq T$, and executes zero or more actions. Action $delay(i)$ sets t_i to $t_i + 1$, action $lengthen(i)$ sets p_i to $p_i + 1$, and action $halt$ halts the alarm clock machine.

Two stipulations are made to guarantee proper operation of an alarm clock machine. First, if its input consists of all "0"'s, then F outputs the null set of actions (the machine is "asleep" until woken). Second, we require that $|p_i/p_j| < O(1)$ for all $1 \leq i, j \leq k$. That is, there is a positive upper bound on the ratio between any two clock periods. This second restriction allows us to more easily simulate our machines. In fact, in our proof of Turing universality, p_i and p_j differ by at most 1. The following theorem asserts the Turing universality of alarm clock machines.

Theorem 18 *There exists an alarm clock machine* (F, k, c) *and a recursive encoding function* $\tau(\mathcal{M})$ *such that for all Turing machines M and binary inputs ω, (F, k, c) halts on input $\langle \tau(\mathcal{M}), \omega \rangle$ if and only if M halts on input ω. Furthermore, if M halts in T steps, then (F, k, c) will halt in $2^{O(T)}$ steps.*

For the proof, we use a series of Turing universal automata: adder machines, acyclic adder machines, and counter machines.

7.1.1   Adder Machines

First, we introduce the adder machines.

Definition 7.1.1 *An adder machine* $\mathcal{D}(k)$ *is a machine consisting of a finite control and k adders, $D_i \ldots D_k$. The operations on the adders are*

- Inc(D_i), which increases D_i by one.

- Compare(D_i, D_j), a function with the range $\{\leq, >\}$.

Definition 7.1.2 An adder machine is said to be *acyclic* if its finite control consists of an acyclic Boolean circuit with no internal memory state.

Lemma 7.1.3 *An adder machine $\mathcal{D}(k)$ with c control states can be simulated by an acyclic adder machine $\mathcal{D}'(k + c)$.*

"*Proof.*" Let $\mathcal{D}$ be an adder machine with c states and k adders. In $\mathcal{D}'$, the state of the finite control is encoded in a "unary representation" via c additional adders: if the current state is i, these c adders will all have the same value, except for the ith adder, whose value is greater by 1. The acyclic circuit reads the "state" from the comparisons between the c adders; it changes "state" from i to j by first incrementing all of the c state adders except for adder i, and then increment adder j. The other k adders are left without change.

We next show the equivalence of adder machines and counter machines, thus proving that adder machines compute all recursive functions.

Definition 7.1.4 *A counter machine $\mathcal{C}(k)$ consists of a finite control and k counters. The counters hold natural numbers; the operations on each counter are as follows: Test for 0, Inc, Dec, and also No change* [HU79].

Lemma 7.1.5 *Adder machines and counter machines are linear time equivalent.*

"*Proof.*"

1. $\mathcal{D}(k) \subseteq \mathcal{C}(O(k^2))$: for each pair i, j of adders in $\mathcal{D}$, $\mathcal{C}$ maintains a pair of counters with the values $\max(0, D_i - D_j)$ and $\max(0, D_j - D_i)$. Comparing between the ith and j^{th} adders is simulated by testing for 0 in both associated counters.

2. $\mathcal{C}(k) \subseteq \mathcal{D}(2k)$: for each counter C_i of $\mathcal{C}$, $\mathcal{D}$ maintains a pair of adders: Inc_i and Dec_i, which are incremented upon $\text{Inc}(C_i)$ and $\text{Dec}(C_i)$, respectively. $\text{Test}(C_i)$ is simulated by comparing Inc_i and Dec_i.

Corollary 7.1.6 *The class of functions computed by an adder machine is recursive. (For any recursive function ψ which is computed by a Turing machine M in time T, there exists an adder machine that computes ψ in time $O(2^T)$.)*

Proof. Counter machines with at least four counters are known to simulate Turing machines with an exponential slowdown ([HU79], page 171, Lemma 7.4). ∎

7.1.2   Alarm Clock and Adder Machines

An alarm clock machine $\mathcal{A}$ can easily be simulated by a counter machine, and hence $\{\mathcal{A}\} \subseteq \{\mathcal{D}\}$. Next, we state the converse inclusion.

Lemma 7.1.7 *Given an acyclic adder machine $\mathcal{D}(k)$ that computes in time T, there exists an alarm clock machine with $O(k^2)$ alarm clocks that simulates $\mathcal{D}$ in time $O(T^3)$.*

The rest of this section is devoted to the proof of lemma 7.1.7.

Proof. Given an acyclic adder machine $\mathcal{D}$ with adders $D_1 \dots D_k$, we construct an alarm clock machine $\mathcal{A}$ with clocks $A_0 \dots A_k$ that simulates $\mathcal{D}$.

The alarm clock A_i simulates the adder D_i. Alarm clock A_0 is used as the 0 value to be compared against by the other k alarm clocks. All of the alarm clocks have the same period, and they differ in the *temporal shift* of their alarming time relative to the alarming of A_0. Hence, if adder D_i is set to n, then A_i alarms n time units after A_0 alarms. It is ensured in the simulation that the period of the clocks is always greater than their phase differences, thus avoiding wraparound problems. The correspondence between adders and alarm clocks is as follows:

Adder$_i$	Alarm clock$_i$
Inc(D_i)	delay(A_i)
Compare(D_i, D_j)	Compare shift phase of clocks A_i and A_j from A_0

One subtlety concerns how to implement the Compare operation. The alarm clock machine's finite control is only allowed to remember the alarm sequence of the previous $O(1)$ time steps. However, after simulating the tth time step of the adder machine, any two alarm clocks may be phase shifted by up to t time units. We need to perform the comparisons and represent this information in a way usable by the short-memory finite control. We accomplish this task by adding a set of $O(k^2)$ auxiliary clocks used to collect the following information:

> For each pair of clocks (i, j), $i < j$, the auxiliary clock A_{ij} determines whether the phase shift of clock A_i is less than or equal to the phase shift of clock A_j. The auxiliary reference clock A_{00} is used to synchronize the auxiliary clocks.

We now describe how the finite control uses the auxiliary clocks to compare the phase shift of the adder alarm clocks. The period of the auxiliary clocks is maintained to be one time unit longer than the period of the adder clocks.

Thus, they alarm one time-step later in each successive cycle of the adder clocks. Conceptually, the finite control uses these auxiliary clocks to sweep through the adder clock cycle, and records the information it needs by delaying the appropriate auxiliary clocks.

Initially, we assume that all of the auxiliary clocks alarm in synchrony with clock A_{00}, and that their phase shift with respect to clock A_0 is less than that of any of the adder clocks (this is easily accomplished by suitably setting the initial conditions). The finite control works as follows:

- If clocks A_{00}, A_{ij} and A_i, but not A_j, alarm simultaneously, then the finite control delays clock A_{ij} once. If A_{00}, A_{ij} and A_j but not A_i alarm, it delays clock A_{ij} twice.

- If clocks A_{00} and A_0 alarm simultaneously, it means that the comparison is done and its result is stored by the auxiliary clocks. The finite control will then be woken up and will receive the results during the next two steps: the alarm pattern of the auxiliary clock A_{ij} determines whether clock A_i's phase shift is less than, equal to, or greater than that of clock A_j, for all clocks A_i, A_j. The finite control then delays the auxiliary clocks so that they will again be synchronous.

It is easy to verify that the controller need only remember the alarm history of the last four time steps in order to perform each of these operations. Once the finite control has the comparison information, it determines if the original adder machine would have halted, and halts accordingly. Otherwise, it determines which adders of the original machine would have incremented, and delays the corresponding clocks. Finally, in order to ensure that the phase shift of the adder clocks does not wrap around, the finite control lengthens the period of all of the clocks by one time unit.

To simulate step t of the adder machine, the alarm clock machine performs the comparisons in time $O(t^2)$ (the period is $O(t)$), and it performs the requisite delays and lengthens the clock periods in $O(1)$ time. Thus, $O(t^3)$ steps are required to simulate t steps of the adder machine, and Lemma 7.1.7 is proven. ∎

7.2 Restless Counters

We now show how to simulate the clocks in a universal alarm clock machine with simple restricted counters, which we call *restless counters*. This will make the simulation of alarm clock machines by sigmoidal networks in the next section easier. We call the event in which the restless counter i is set to "0" a *zero event for* i, and the event in which some restless counter is set

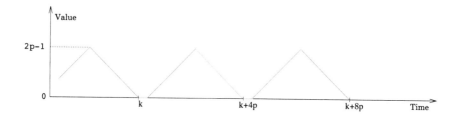

Figure 7.2: The values of a restless counter associated with an alarm clock of period p

to "0" a *zero event*. The resulting restless counter machine has the following properties:

- Every restless counter must always either be incremented or decremented in each time step, unless it is at "0", in which case it must be incremented within $O(1)$ steps.

- Once a restless counter is instructed by the finite control to start incrementing (decrementing), it must continue to do so in each successive time step until it has an opportunity to change its direction. (The finite control can "sleep" after giving the command of what direction to continue with.)

- A clock alarming must correspond to a zero event.

- The finite control is allowed to change the directions of the restless counters only during $O(1)$ time steps following a zero event (because the control can only be awakened by a zero event).

- At any time step, for any i, there will be at most $O(1)$ zero events before the next zero event for i.

We next show our particular implementation. To simplify matters, we assume that the universal alarm clock machine runs a valid simulation of an acyclic adder machine, and thus behaves as described in the previous section.

We implement each alarm clock i with a pair of restless counters, which we call *morning* (M_i) and *evening* (E_i). When the clock is in its steady state (neither delayed or lengthened) with period p, each restless counter has the periodic behavior described in Figure 7.2.

That is, it counts up to $2p - 1$, then down to "0", stays at "0" for two time steps and starts counting up again:

$$\cdots 0\ 0\ 1\ 2\ 3\ 4\ \cdots\ (2p - 1)\ \cdots 4\ 3\ 2\ 1\ 0\ 0\ 1\ 2\ 3 \cdots$$

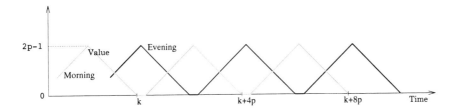

Figure 7.3: A pair of counters simulating an alarm clock of period p, in steady state after alarming at time k

M :	1←0	0←0	0←0	0←1	1←2	2←3	3←4	4←5	5←4	4←3	3←2	2←1	1←0	0←0
E :	3←4	4←5	5←4	4←3	3←2	2←1	1←0	0←0	0←0	0←1	1←2	2←3	3←4	4←5

$$k \qquad\qquad\qquad\qquad\qquad\qquad\qquad\qquad\qquad\qquad k+12$$

To achieve this oscillatory effect, we put M_i and E_i $2p$ time steps out of phase. If M_i (resp. E_i) is decremented to 0 at time k, then E_i (resp. M_i) (which has been incrementing) starts decrementing at time $k+1$ and M_i (resp. E_i) starts incrementing at time $k+2$.

Thus, in its steady state, the system oscillates with a period of $4p$. We identify a single time unit of an alarm clock with four time units of the restless counter, and interpret the event of the restless counter M_i turning from 1 to 0 with the clock alarming, (see Figure 7.3). We refer to the time gap between two successive alarming of a morning restless counter as a day. The operations *delay* and *lengthen* (to be described next) can be referred to as delaying the next day or lengthening the duration of all days from now on.

(This construction does not handle clocks with period 1. However, such clocks are not necessary for our alarm clock machine to be universal.) Here is a numerical example for $p = 3$:

We now show how to implement the *delay* and *lengthen* operations. For these operations, we assume that neither of the restless counters is equal to 0, and that it is known which restless counter is decrementing and which is incrementing. Upon inspection of our "program", one can verify that the finite control will always have this information within $O(1)$ time units after it has woken up, and that it must wait only $O(1)$ steps before the nonzero condition is met. For example, when the finite control has received all of its comparison information, it can wait a few steps and ensure that the morning restless counters of all the comparison clocks and the 0 clock are incrementing, while the evening restless counters of all the adder clocks are decrementing.

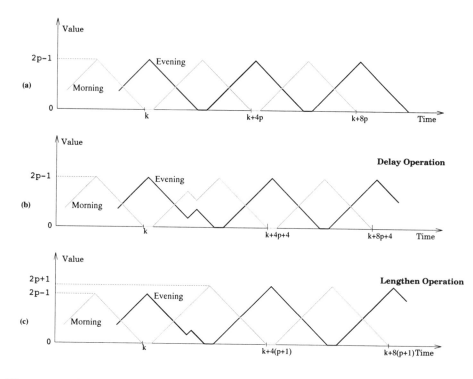

Figure 7.4: A pair of Restless Counters (a) simulating an alarm clock in steady state (b) simulating a delay in alarming (c) simulating lengthening the period of an alarm clock.

To delay a clock, the finite control increments the restless counter it had previously been decrementing, and decrements the restless counter it had previously been incrementing both a duration of for two time steps, and then resumes its normal operations, (see Figure 7.4).

To lengthen the day's period, the finite control increments for one time step the restless counter that it had previously been decrementing, without changing the direction of the other counter, and then continues with normal operation. Note that this operation will also alter the phase shift of the restless counters. However, since the operation will be performed on all of the clocks in the simulation, the relative phase shifts will be preserved.

7.3 Sigmoidal Networks are Universal

In this section, we prove the main theorem of the chapter:

Theorem 19 *Given an alarm clock machine $\mathcal{A}$ (with no input) that simulates* *an acyclic adder machine, there is a ϱ-network $\mathcal{N}$ that computes the same function as $\mathcal{A}$ and requires the same computation time. Furthermore, the size of this network is linear in the number of alarm clocks, and in the size of the finite control of $\mathcal{A}$.*

The rest of this section is dedicated to the proof of Theorem 19. We first state four properties of our sigmoid (and "similar" functions) that are used in the proof:

- **Feature 1:** There exists a positive constant c, such that $\forall x, |x| > c$, $\varrho(x)$ is monotonically nondecreasing, and $\varrho(x) \in [-1, \frac{-1}{2}]$ or $\varrho(x) \in [\frac{1}{2}, 1]$, depending on the sign of x.

- **Feature 2:** For every constant b, for which the slope of $\varrho(bx)$ at 0 is larger than 1, $\varrho(bx)$ has three fixed points. One is at zero and the two others are at A and $-A$, for a constant A. (For example, for the sigmoid $\varrho(x) = \frac{2}{1+e^{-x}} - 1$, the slope at 0 is $\frac{1}{2}$ and this feature requires $b > 2$.) The larger b is, the closer A gets to 1 $(0.5 < A < 1)$. For example, using 15 decimal digits in the precision:

$$b = 5 \quad A = 0.98562369130483$$
$$b = 10 \quad A = 0.999909121699349$$
$$b = 30 \quad A = 0.999999999999812 \,.$$

Let c be a constant. In the equation $\varrho(bx + c)$, the two external fixed points are not equal in size anymore, provided that $c \neq 0$, and they thus are denoted by A_1 (≈ -1) and A_2 (≈ 1).

The fixed points A_1 and A_2 are stable and constitute exponential attractors (for all $x \neq 0$), and the middle fixed point $x = 0$ is unstable. (In fact, one can achieve d^{-t} convergence for any d, $0 < d < 1$ by a suitable choice of the constant b.)

- **Feature 3:** The function ϱ is differentiable twice around its attractors.

- **Feature 4:** For every x, $\varrho(x) = x + O(x^3)$. (Hence, if x is a small number, then $\varrho(x) \sim x$.) This is proved for the sigmoid by a Taylor expansion around 0.

We will use the above four properties of our sigmoid to prove its universality. Given an alarm clock machine $\mathcal{A}$ (with restless counters), the network $\mathcal{N}$ that simulates $\mathcal{A}$ consists of three main components: a finite control, a set of restless counters, and a set of flip-flops. The finite control and the restless counter parts of the network simulate the corresponding components of the alarm clock machine. Since the finite control is memoryless, we need a third mechanism for controlling the restless counters. This is accomplished through

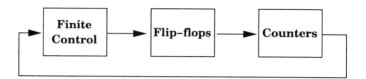

Figure 7.5: Block Diagram of the Sigmoidal Network

the implementation of a set of two-state flip-flop neurons that serve as inter-mediaries between the finite control and the restless counters.

Implementing the Finite Control

It has long been known how to simulate any finite control of $\mathcal{A}$, $FC_{\mathcal{A}}$, by a network of threshold devices (see Subsection 1.3.1). If the original finite control has finite memory, i.e. it depends only on the last $O(1)$ time steps, the resulting threshold network can be made into a feed-forward one.

We substitute each threshold device

$$x_i(t+1) = \mathcal{H}(\sum_{i=1}^{N} a_{ij}x_j + \theta_i)$$

with a sigmoidal device

$$x_i(t+1) = \varrho(\alpha_a(\sum_{i=1}^{N} a_{ij}x_j + \theta_i)) \,,$$

for a large fixed constant α_a. As long as the summation in the above expression is guaranteed to be bounded away from 0, the output values of the neuron using the sigmoid activation function will closely approximate the output of the neurons using the Heaviside activation function. By choosing α_a sufficiently large, we can make this approximation as close as we desire (feature 1).

Note that the number of states in our "finite control" is in fact infinite, because every neuron can take on an infinite set of activation values. Because these values fall within a small neighborhood of either "1" or "-1", we can conceptually discretize them; however, the continuous nature of these values leads to accuracy problems.

For each restless counter i, the finite control has two output lines (imple-mented as neurons), *Start-Inc$_i$* and *Start-Dec$_i$*. When *Start-Inc$_i$* is active (i.e., $\approx$ "1"), it means that restless counter i should be continually incremented.

Similarly, when $Start\text{-}Dec_i$ is active, it means that restless counter i should be continually decremented. Most of the time, both output lines are in an inactive state (i.e., $\approx$ "0"). In this case, restless counter i is handled according to the last issued command, allowing operations to be performed on it when the finite control is inactive. It will never be the case that both signals are simultaneously active.

Bi-directional Flip-flops

Recall that to avoid unrecoverable data corruption, we implement a "finite control" that converges to a constant "ground state" during the long periods between interesting events. In order to maintain control of the restless counters during these quiet periods, we introduce special flip-flop devices. These devices will have two stable states, and are guaranteed to exponentially converge to one of them during the quiet periods. When the finite control is active, it can set or reset the value of a flip-flop. Otherwise, the flip-flop maintains its current state.

The update equation of each flip-flop is

$$\text{ff}_i = \varrho(\alpha_{f1}(\text{ Start-Inc}_i - \text{Start-Dec}_i) + \alpha_{f2}\text{ff}_i + \alpha_{f3}) ,$$

where α_{f1}, α_{f2} and α_{f3} are suitably chosen constants (feature 2).

Counters

Each restless counter is implemented by three sigmoidal neurons: one, called the rc neuron, retains the value of the restless counter, and the other two assist in executing the Inc/Dec operations. A restless counter with the value $v \in \mathbb{N}$ is implemented in an rc neuron with a value "close" to B^{-v}, where B is a constant, $B > 2$. That is, a value "0" in a restless counter is implemented as a constant close to "1" in the neuron. When a restless counter increases, the associated rc neuron decreases by a factor of B.

Thus, at each step, the rc neuron is multiplied by either B or $\frac{1}{B}$. To do this, we use the approximation:

$$\varrho(V + cx_i) - \varrho(V) \approx \varrho'(V)cx_i ,$$

for sufficiently small c and $|x_i| < 1$ (feature 3). Let V be the "direction" input signal, coming from the ith flip flop. That is, V converges to either A_1 or A_2. An rc neuron updates itself by the equation:

$$
\begin{aligned}
x_i(t+1) &= \varrho[\alpha_{c1}\varrho(\alpha_{c2}V + \alpha_{c3} + \alpha_{c4}x_i(t)) - \alpha_{c1}\varrho(\alpha_{c2}V + \alpha_{c3}) \\
&\quad + \alpha_{c5}x_i(t)] \\
&\approx \varrho[(\alpha_{c1}\varrho'(\alpha_{c2}V + \alpha_{c3}))\alpha_{c4}x_i(t) + \alpha_{c5}x_i(t)]
\end{aligned}
$$

By a suitable choice of the constants $\alpha_{c1}, \ldots, \alpha_{c5}$, we have:

$$\alpha_{c1}\alpha_{c4}\varrho'(\alpha_{c2}A_1 + \alpha_{c3}) + \alpha_{c5} = B$$
$$\alpha_{c1}\alpha_{c4}\varrho'(\alpha_{c2}A_2 + \alpha_{c3}) + \alpha_{c5} = \tfrac{1}{B} .$$

If the value of x_i is close enough to 0, we can approximate (using feature 4)

$$\varrho[(\alpha_{c1}\varrho'(\alpha_{c2}V + \alpha_{c3}) + \alpha_{c4})x_i] \approx (\alpha_{c1}\varrho'(\alpha_{c2}V + \alpha_{c3}) + \alpha_{c4})x_i .$$

This discussion provides the intuition of why the rc neuron computes either $\sim Bx_i$ or $\sim \frac{1}{B}x_i$. Note also that when x_i is positive and "close to 1," and it is "multiplied by B" then it will in fact be drawn towards a fixed point of the above equation. This acts as a form of *error correction*.

7.3.1   Correctness of the Simulation

To avoid heavy notation, we will sketch the proof to a point at which we believe it will be clear enough to enable the reader to fill in the details. Ideally, our finite-state neurons would all have $\{0,1\}$ values, our flip-flops would take on precisely two values (A_1, A_2) and the rc neuron would have the exact activation B^{-v}, where v is the value of the simulated restless counter. Unfortunately, it is inevitable that the neurons' values will deviate from their ideal values. To obtain our result, we show that these errors are controllable.

The proof of the correctness of the simulation is inductive on the serial number of the day (that is, the times of M_0 alarming). As the network $\mathcal{N}$ consists of three parts: finite automaton (FA), flip flops (FF), and restless counters, for each part we assume a "well behaved input" in day d and prove a "well behaved output" for the same day. Because on the first day inputs to all parts are well-behaved, the correctness follows inductively.

Lemma 7.3.1 *Each of the claims stated below implies the next claim in cyclic order:*

(1) On each day d, FC sends $O(1)$ signals (intentionally non-zero) to the flip flops. Each signal has an error bounded by $\mu < .01$. The sum of errors in the signals of the FC during the dth day is bounded by the constant $\beta < 0.1$.

(2) On each day d, $O(1)$ of the signals sent by FF have an error of γ, where γ can be made arbitrarily small (as a function of μ and the constants of the flip flops). The sum of the total error of all signals during the dth day is bounded by δ, where δ can also be made arbitrarily small.

(3) On each day d, a restless counter with a value y acquires a total multiplicative error $\zeta < 0.01$. That is, the ratio of the actual value with the ideal value will always be between 0.99 and 1.01.

Proof.
1 ⇒ 2:
Assume that the finite control sends *Start-Inc$_i$* and *Start-Dec$_i$* to ff$_i$, and that both these two values are never active simultaneously. The update equation of each flip-flop is

$$\text{ff}_i = \varrho(\alpha_{f1}(\text{ Start-Inc}_i - \text{Start-Dec}_i) + \alpha_{f2}\text{ff}_i + \alpha_{f3}) \ .$$

- When either *Start-Inc$_i$* or *Start-Dec$_i$* is active, ff$_i$ is set to the new value. The error γ is bounded by

$$\gamma \leq |1 - \varrho(\alpha_{f1}(1 - \mu) - \alpha_{f2} + \alpha_{f3})| \ .$$

It is easy to see that when $|\alpha_{f1}| - |\alpha_{f2}| + |\alpha_{f3}|$ increases, γ decreases. That is, γ is controllable. For example, if $\alpha_{f1} \geq \mu^{-1}$, α_{f2}, and $\alpha_{f3} \leq 20$ then $\gamma < 0.01$.

- When both *Start-Inc$_i$* and *Start-Dec$_i$* are small, ff$_i$ converges to its closest fixed point. If (*Start-Inc$_i$* − *Start-Dec$_i$*) were exactly "0", then ff$_i$ would be attracted at an exponential rate to its closest fixed point. If α_{f2} is large enough, the fixed points can be made arbitrarily close to "-1" and "1". Furthermore, noise from $\alpha_{f1}(\text{Start-Inc}_i - \text{Start-Dec}_i)$ can be arbitrarily attenuated, since $|\alpha_{f1}\varrho'(\alpha_{f2}\text{ff}_i + \alpha_{f3})|$ can be made vanishingly small by a suitable choice of constants.

2 ⇒ 3:
The update equation of a restless counter x_i is given by

$$x_i = \varrho[\alpha_{c1}\varrho(\alpha_{c2}V + \alpha_{c3} + \alpha_{c4}x_i) - \alpha_{c1}\varrho(\alpha_{c2}V + \alpha_{c3}) + \alpha_{c5}x_i] \ .$$

We next show that by using such an update equation, the rc neuron x_i multiplies itself in each step by either $\sim B$ or $\sim \frac{1}{B}$, allowing small controllable error. Recall that for small y,

$$\varrho(\varrho(V + y) - \varrho(V)) \approx \varrho'(V)y.$$

We can choose constants $\alpha_{c1}, \alpha_{c2}, \alpha_{c3}, \alpha_{c4}, \alpha_{c5}$ such that

$$\alpha_{c1}\alpha_{c4}\varrho'(\alpha_{c2}A_1 + \alpha_{c3}) + \alpha_{c5} = B$$
$$\alpha_{c1}\alpha_{c4}\varrho'(\alpha_{c2}A_2 + \alpha_{c3}) + \alpha_{c5} = \frac{1}{B} \ .$$

The deviation from this ideal behavior is caused by three elements:

1. the error caused by approximating the difference equation by the differential,

2. the error in $\varrho'(\alpha_{c2}V + \alpha_{c3})$ relative to the desired $\varrho'(\alpha_{c2}A_i + \alpha_{c3})$,

3. the error caused by using the approximation $\varrho(x) \approx x$ for small x.

In the first case, the multiplicative error is proportional to $\varrho''(A_i)(\alpha_{c4}x_i)^2$. However, x_i shrinks exponentially (and then grows back in a symmetric manner). Hence, on a given day, these terms form two exponentially decreasing sums. In the second case, we can bound the resulting multiplicative error by a function of $\alpha_{c1}, \alpha_{c2}, \alpha_{c4}$ and $\varrho''(\alpha_{c2}A_i + \alpha_{c3})$ times the error in V relative to A_i. Finally, note that $\varrho(x) = x + O(x^3) = x(1 + O(x^2))$. Since x_i exponentially vanishes (and reappears), the multiplicative error terms form two exponentially decreasing sums.

By "summing" all of these multiplicative errors, we get the desired bound. We can then use the identity that

$$(1 + \delta_1)(1 + \delta_2) \cdots (1 + \delta_k) = 1 + O(\delta_1 + \cdots + \delta_k)$$

when the sum $\delta_1 + \cdots + \delta_k$ is sufficiently small, to approximate the multiplicative error.

$3 \Rightarrow 1$:
Because the finite control is feedforward, and since each restless counter alarms $O(1)$ times a day, the finite control will output (intentionally) nonzero signals only $O(1)$ times a day. By adjusting the constants in our implementation of the finite control, we can make them have arbitrarily small errors when they change their values. During the quiescent period, if the restless counters were actually at 0, then the *Start-Inc$_i$* and *Start-Dec$_i$* neurons would converge exponentially to some canonical value, and their difference would converge exponentially close to 0. We can bound the errors caused by the restless counters being nonzero by some constant c times the sum of the values of all the restless counters at every time of the day. By choosing the weights appropriately, we can, in fact, make c as small as desired.   ∎

7.4   Conclusions

The proof of universality is not limited to the particular sigmoid of Equation (7.1), but rather can be generalized. Let $\tilde{\varrho}$ be any function that adheres to features 1–4, as defined in Section 7.3. We call all such functions by the name *sigmoid-like* functions. The proof of Theorem 19 can be generalized to any sigmoidal-like network.

Corollary 7.4.1 *Let $\tilde{\varrho}$ be any sigmoid-like function. Given an alarm clock machine $\mathcal{A}$ (with no input) that computes a function ψ in time T, there is* a $\tilde{\varrho}$-network $\mathcal{N}$ that computes ψ in time $O(T)$. Furthermore, the size of this network is linear in the number of clocks and the size of the finite control of $\mathcal{A}$.

We conclude that Turing universality is a general property of many recurrent neural networks. Implications of this result include: there is no computable limit on the running time of a network; it is not possible to determine whether a network ever converges or enters a detectable oscillatory state, or even whether a given neuron ever gets close to 0 or 1. Please refer to Chapter 3 for an elaborated discussion on the implications of undecidability.

Chapter 8

Different-limits Networks

In Chapter 7 we showed that the saturated-linear activation function is not unique in its Turing universality, but rather that various sigmoidal-like activation functions can form finite-size architectures which are Turing universal as well. The class of activation functions considered in this chapter is much wider than that of the previous chapter, and as a result the lower bound on its computational power is weaker. We prove that any function for which the left and right limits exist and are different can serve as an activation function for the neurons to yield a network that is at least as strong computationally as a finite automaton. For the exact statement of this observation we need the following definition:

Definition 8.0.2 A function ϱ is said to be of *different-limits* if both $\lim_{x \to \infty} \varrho(x) = t_+$ and $\lim_{x \to -\infty} \varrho(x) = t_-$ exist and $t_+ \neq t_-$.

Let ϱ be any different-limits function. We show that any network of ϱ-neurons of the type

$$x_i(t+1) = \varrho \left(\sum_{j=1}^{N} a_{ij} x_j(t) + \sum_{j=1}^{M} b_{ij} u_j(t) + c_i \right) , \quad i = 1, \ldots, N \qquad (8.1)$$

can simulate a finite automaton. Thus, any such network can be chosen for control implementation, depending only on properties such as the cost, availability, and ease of learning of the particular application. Constructing optimal neural networks for given regular languages is an important topic for applications, but is outside the scope of this book; see e.g., [OG96, SG97].

8.1 At Least Finite Automata

The main theorem of this chapter is as follows.

Theorem 20 *Let $\mathcal{A}$ be a finite automaton. Then for every different-limits function ϱ there exists a neural network which simulates $\mathcal{A}$ with ϱ as an activation function.*

To prove this theorem, we first state the following interpolation fact.

Lemma 8.1.1 *For any different-limits activation function ϱ, there exist constants $a_0, a_i, b_i, c_i \in \mathbb{R}$, $i = 1, 2, 3$ such that the function*

$$f(y) = a_0 + \sum_{i=1}^{3} a_i \varrho(b_i y + c_i)$$

satisfies $f(-1) = f(0) = 0$ and $f(1) = 1$.

We postpone the proof of this lemma to the next section; here we show how it leads to theorem 20.

We prove that any finite automaton with s states and m input values can be simulated by a network of $N = 3sm$ neurons whose activation function has different-limits.

Proof. (*Of Theorem 20*)
We employ the general definition of an automaton with no initial state and with sequential output, as described in Section 1.2. Suppose that we are given two automata $\mathcal{M} = (Q, I, Y, f, h)$ and $\overline{M} = (Q, I, Y, \overline{f}, \overline{h})$ that have the same input $(I = \Sigma \bigcup \{\$\})$ and output (Y) sets, and where f^* and $\overline{f}^*$ are the generalized transition functions as described in Subsection 1.3. The automaton $\overline{\mathcal{M}}$ *simulates* $\mathcal{M}$ if there exist two maps

$$\text{ENC} : Q \to \overline{Q} \quad \text{and} \quad \text{DEC} : \overline{Q} \to Q \,,$$

called the *encoding* and *decoding* maps respectively, such that for each $q \in Q$ and each sequence $u \in \Sigma^*$,

$$f^*(q, u) = \text{DEC}\left[\overline{f}^*(\text{ENC}\,[q], u)\right], \quad h(q) = \overline{h}(\text{ENC}\,[q]) \,.$$

Assume that $\Sigma = \{e_1, \ldots, e_m \in \mathbb{R}^m\}$, where e_i is the ith canonical basis vector, i.e., the vector having a "1" in the ith position and "0" in all other entries. Similarly, suppose that $Y = \{e_1, \ldots, e_\ell \in \mathbb{R}^\ell\}$, and the 0 vectors denote no-information in both input and output ends. (The assumption that Σ and Y are of this special "unary" form is not very restrictive, as one may always encode inputs and outputs in this fashion.) We next think of u_v as the predicate "current input is e_v" and of y_l as "current output is e_l." We also let $Q = \{q_1, \ldots, q_s\}$.

We next construct a neural network $\mathcal{N}$ that simulates $\mathcal{M}$. $\mathcal{N}$ has $N = 3sm$ neurons, which we denote by x_{ijk}, where $i = 1, \ldots s$, $j = 1, \ldots, m$, and

$k = 1, 2, 3$. For the construction, we utilize the Boolean variables g_{rv} for $r = 1, \ldots, s$, and $v = 1, \ldots, m$, that indicate if the current state of $\mathcal{M}$ is q_r and the last input read was e_v.

$$g_{rv} = a_{rv} + \sum_{k=1}^{3} a_{rvk}\, x_{rvk} , \qquad (8.2)$$

where the constants a_{rv}, a_{rvk}, will be characterized later. In terms of these quantities, the update equations of the neurons for $r = 1, \ldots, s$, $v = 1, \ldots, m$, $k = 1, 2, 3$, can be expressed by:

$$x_{rvk}^{+} = \varrho \left(b_{rvk}(\sum_{j=1}^{m} \sum_{i \in Q_{rv}} g_{ij} + u_v - 1) + c_{rvk} \right) , \qquad (8.3)$$

where

$$Q_{rv} := \{ i \mid f(q_i, e_v) = q_r \},$$

g_{rv} is the "macro" of Equation (8.2), and b_{rvk} as well as c_{rvk} will be specified below. Finally, for each $l = 1, \ldots, \ell$, the lth coordinate of the output is defined by:

$$y_l = \varrho \left(c \sum_{j=1}^{m} \sum_{i \in T_l} g_{ij} \right) \qquad (8.4)$$

where $T_l := \{ i \mid h_l(q_i) = 1 \}$ for the coordinate h_l of h, and c is any constant such that $\varrho(0) \neq \varrho(c)$.

The proof that this is indeed a simulation is as follows:

1. We think of g_{rv} as the entries of a matrix $[g]_{rv} \in \mathbb{R}^{sm}$. We first prove inductively on the steps of the algorithm that the matrices $[g]_{rv}$ are always of the type E_{ij}, where E_{ij} denotes the binary matrix whose ij^{th} entry has the value "1" and all the rest have the value "0". Furthermore, except for the starting time, E_{rv} indicates that the simulated finite automata is in state r and its last read input was v. As a result, the parameter multiplying b_{rvk} in Equation (8.3) may only admit values in $\{-1, 0, 1\}$.

   - We start the network with an initial state $x_0 \in \mathbb{R}^N$ so that the starting $[g]_{ij}$ has the form E_{r1}, where q_r is the corresponding state of the original automaton. It is easy to verify that such x's are possible, since in Equation (8.2) the different equations are uncoupled for different r and v, and not all weights a_{rvk} for $k = 1, 2, 3$ can vanish, otherwise in Lemma 8.1.1 the function f (which is used as g_{rv}) would be constant.

- Assume that at some time t, only one of the g_{rv} has the value "1" and the rest are "0", the associated state of the finite automata is q_r, and the last read input is e_v. Then, the expression

$$y_{rv} = \left(\sum_{j=1}^{m} \sum_{i \in S_{rv}} g_{ij} + u_v - 1 \right)$$

can only take the values "-1", "0", or "1". The value "1" can only be achieved for this sum if both $u_v = 1$ and there is some $i \in S_{rv}$ so that $g_{ij} = 1$, that is, if the current state of the original machine is q_i and $f(q_i, e_v) = q_r$.

By Lemma 8.1.1, there exist values $a_{rv}, a_{rv}, a_{rvk}, b_{rvk}, c_{rvk}$ ($k = 1, 2, 3$) so that $f(y_{rv})$ (which is by definition here the value of g_{rv} at time $t+1$) assumes the value "1" only if y_{rv} was "1", and is "0" for the other two cases. This proves the correctness in terms of the expressions $[g]_{ij}$. The vectors x_{rvk} take the values $\varrho(b_{rvk} y_{rv} + c_{rvk})$, as described by Equation (8.3).

2. The encoding and decoding functions are defined as follows: The encoding map ENC $[q_r]$ maps q_r into any fixed vector x so that Equation (8.2) gives $[g]_{ij} = E_{r1}$. The decoding map DEC $[x]$ maps those vectors x that result in $[g]_{ij} = E_{rv}$ ($r = 1 \ldots s$, $v = 1 \ldots m$) into q_r, and is arbitrary on all other elements of $\mathbb{R}^N$.

∎

In the next section, we prove the lemma.

8.2   Proof of the Interpolation Lemma

Proof. We prove more than the actual statement, namely, we show that for each choice of three numbers $r_{-1}, r_0, r_1 \in \mathbb{R}$ there exist $a_0, a_i, b_i, c_i \in \mathbb{R}$, $i = 1, 2, 3$ so that, denoting

$$f(y) = a_0 + \sum_{i=1}^{3} a_i \varrho(b_i y + c_i) \,,$$

it holds that $f(-1) = r_{-1}$, $f(0) = r_0$, and $f(1) = r_1$. To prove this, it suffices to show that there are b_i, c_i ($i = 1, 2, 3$), so that the matrix

$$\Gamma_{bc} = \begin{pmatrix} \tilde{\varrho}[b_1(-1) + c_1] & \tilde{\varrho}[b_2(-1) + c_2] & \tilde{\varrho}[b_3(-1) + c_3] \\ \tilde{\varrho}[b_1(0) + c_1] & \tilde{\varrho}[b_2(0) + c_2] & \tilde{\varrho}[b_3(0) + c_3] \\ \tilde{\varrho}[b_1(1) + c_1] & \tilde{\varrho}[b_2(1) + c_2] & \tilde{\varrho}[b_3(1) + c_3] \end{pmatrix} \qquad (8.5)$$

where $\tilde{\varrho}$ is a certain function of the form $a\varrho(x) + b$, is non-singular. Hence, for all $R = \mathrm{COL}\,(r_{-1}, r_0, r_1)$ there exists a vector $A = \mathrm{COL}\,(a_1, a_2, a_3)$ so that

$$\Gamma_{bc}A = R.$$

Let m_i, $i = 1, 2, 3$ be maps on the real numbers so $(m_i(-1), m_i(0), m_i(1)) = e_i$ ($e_i \in \mathbb{R}^3$ is the ith canonical vector). Let $U = \{-1, 0, 1\}$. We say that k ϱ-neurons g_j *linearly interpolate[U] the map* m_i if there exist constants $a_1^i, \ldots, a_k^i$ so that $f_i(u) = \sum_{j=1}^{k} a_j^i g_j(u)$ and

$$f_i(u) = m_i(u)$$

for all $u \in U$. These neurons are said to ϵ-*approximate[U]* m_i if

$$|f_i(u) - m_i(u)| < \epsilon$$

for all $u \in U$.

Proposition 8.2.1 *There are three $\mathcal{H}$-neurons (see Subsection 1.3.1) that interpolate[U] the maps m_i, $i = 1, 2, 3$.*

Proof. Let

$$
\begin{aligned}
h_1(x) &= \mathcal{H}(x + \tfrac{1}{2}) \\
h_2(x) &= \mathcal{H}(x - \tfrac{1}{2}) \\
h_3(x) &= \mathcal{H}(-x - \tfrac{1}{2}) .
\end{aligned}
$$

The interpolation is by:
$m_1 = h_3$ ($a_1^3 = 0, a_2^3 = 0, a_3^3 = 1$),
$m_2 = h_1 - h_2$ ($a_1^2 = 1, a_2^2 = -1, a_3^2 = 0$), and
$m_3 = h_2$ ($a_1^1 = 0, a_2^1 = 1, a_3^1 = 0$). ∎

Proposition 8.2.2 *For all $\epsilon > 0$ there are three ϱ-neurons that ϵ-interpolate[U] the maps m_i, $i = 1, 2, 3$.*

Proof. Because ϱ has two different limits, we can impose $t_+ = 1$, and $t_- = 0$ on ϱ without restricting the affine span of the neurons. This is possible by defining the function

$$\tilde{\varrho} = \frac{\varrho(x) - t_-}{t^+ - t_-} . \tag{8.6}$$

Without loss of generality, we assume $t^+ = 1$ and $t_- = 0$ from now on. So, for each $\epsilon > 0$ there is some $\eta > 0$ such that for all y, $|y - \tfrac{1}{2}| > \eta$

$$|\varrho(y - \tfrac{1}{2}) - \mathcal{H}(y - \tfrac{1}{2})| < \epsilon .$$

In particular, $\forall y, |y - \frac{1}{2}| > \frac{1}{4}$ we can choose $\lambda > 4\eta$, and by using the relation $\mathcal{H}(\lambda(y - \frac{1}{2})) = \mathcal{H}(y - \frac{1}{2})$, we obtain

$$|\varrho(\lambda(y - \frac{1}{2})) - \mathcal{H}(y - \frac{1}{2})| < \epsilon \ .$$

A similar argument can be applied to $(y + \frac{1}{2})$ and $(-y - \frac{1}{2})$. We conclude that for all $\epsilon > 0$ there exists some λ so that

$$
\begin{aligned}
g_1(x) &= \varrho(\lambda(x + \frac{1}{2})) \\
g_2(x) &= \varrho(\lambda(x - \frac{1}{2})) \\
g_3(x) &= \varrho(\lambda(-x - \frac{1}{2}))
\end{aligned}
$$

satisfy $|g_i(x) - h_i(x)| < \frac{\epsilon}{3}$ for all $u \in U$. The result is now clear. ∎

From Propositions 8.2.1 and 8.2.2 we conclude that for all $\epsilon > 0$ there are g_1, g_2, g_3 so that

$$\left| \Gamma_{bc} \begin{pmatrix} 0 & 1 & 0 \\ 0 & -1 & 1 \\ 1 & 0 & 0 \end{pmatrix} - \begin{pmatrix} 1 & 0 & 0 \\ 0 & 1 & 0 \\ 0 & 0 & 1 \end{pmatrix} \right| < \epsilon,$$

where Γ_{bc} is the matrix defined in Equation (8.5) and $\tilde{\varrho}$ is related to ϱ by Equation (8.6). Choosing ϵ so that Γ_{bc} is non-singular, the lemma results. ∎

To summarize, we have proved a lower bound on the computational power of a large class of recurrent neural networks whose activation function is only constrained to have different limits at $\pm\infty$.

Chapter 9

Stochastic Dynamics

Having understood the power of deterministic analog recurrent neural networks, we now turn our attention to networks that exhibit stochastic and random behavior. Randomness is a basic characteristic of large distributed systems. It may result from the activity of the individual agents, from unpredictable changes in the communication pattern among the agents, or even just from the different update paces. All previous work that examined stochasticity in networks, e.g., [vN56, Pip90, Adl78, Pip88, Pip89, DO77a, DO77b], studied only acyclic architectures of binary gates, while we study general architectures of analog components. Due to these two qualitative differences, our results are totally different from the previous ones, and require new proof techniques.

Our particular stochastic model can be seen as a generalization of the von Neumann model of unreliable interconnections of components, where each basic component x_i has a fixed probability p_i for malfunction at any step [vN56]. In contrast to the von Neumann model, here it is natural to allow for real values in p_i, rather than rational values only. Furthermore, p_i can be either a constant, as in the von Neumann model, or alternatively, a function of the history and the neighboring neurons. The latter, referred to as "the Markovian model," provides a useful model for stochastic computation. The element of stochasticity, when joined with exact known parameters, has the potential to increase the computational power of the underlying deterministic process. We find that it indeed adds some power, but only if the weights are rationals. In the cases of real weights and integer weights, this type of stochasticity does not change the computational power of the underlying process.

The proof concerning rational weights includes the following result from the realm of theoretical computer science. It is well known that probabilistic Turing machines that use binary coins with rational probabilities compute the class BPP. Here we consider binary coins having *real* probabilities and prove that the resulting polynomial time computational class is BPP/ log $*$, which is BPP augmented with prefix logarithmic advice.

It is perhaps surprising that the real probabilities strengthen the Turing machine, because the machine still reads only the binary values of the coin flips. However, a long sequence of coin flips allows indirect access to the real valued probability, or more accurately, it facilitates its approximation with high probability. This is in contrast to the case of real weight networks, where access to the real values is direct and immediate. Thus, the resulting computation class $(\mathrm{BPP}/\log *)$ is of intermediate computational power. It contains some nonrecursive functions, but is strictly weaker than P/poly.

Because real probabilities do not provide the same power as real weights, this chapter can be seen as suggesting a model of computation that is stronger than a Turing machine, but still is not as strong as real weight neural networks. Complementary to the feature of "linear precision suffices" for real weights (Chapter 4) we prove that for stochastic networks "logarithmic precision suffices" for the real probabilities; that is, for up to the qth step of the computation, only the first $O(\log q)$ bits in the probabilities of the neurons influence the result. We note that the same precision characterizes the quantum computer.

The notion of stochastic networks is closely related to the concept of deterministic networks influenced by external noise. A recent series of works considers recurrent networks in which each neuron is affected by additive analog noise; the noise is described as a continuous random variable, characterized by a density function. Typically, the resulting networks are not stronger than finite automata [Cas96, OM98], and for many types of noise, they compute a strict subset of the regular languages [MS98, Rab63, SR98]. In the stochastic networks considered in this chapter the noise disturbance of a neuron is a discrete variable characterized by a probability function which is non-zero in no more than a few domain points. The applicability of such stochasticity to part of the network can only increase the computational power. Therefore, the term "noisy" is not a proper description of our network, and we prefer the term "stochastic."

This chapter is organized as follows: Section 9.1 focuses on our stochastic networks, distinguishing them from a variety of stochastic models. Section 9.2 states the main results. Sections 9.3-9.5 include the proofs of the main theorems. In Section 9.6 we restate the model in various forms and in Section 9.7 we briefly describe a particular form of nondeterministic stochastic networks.

9.1   Stochastic Networks

Four main questions are to be addressed when considering stochastic networks. How do we model stochasticity? What type of random behavior (or errors) should be allowed? How much randomness can be handled by the model? Finally, stochastic networks are not guaranteed to generate the same response in different runs of a given input; thus, how do we define the output of the network for a given input?

Modeling Stochasticity

The first question, how to model stochasticity, was discussed by von-Neumann [vN56] and quoted by Pippenger [Pip90]:

> "The simplest assumption concerning errors is this: With every basic organ is associated a positive number ϵ such that in any operation, the organ will fail to function correctly with the (precise) probability ϵ. This malfunctioning is assumed to occur statistically independently of the general state of the network and of the occurrence of other malfunctions."

We first adopt von Neumann's statistical independence assumption. This assumption is also consistent with the works of Wiener [Wie49] and Shannon [Sha48], where the "noise"–which is the source of stochasticity–is modeled as a random process with known parameters. Note that in this model, the components are stochastic in precisely the amount p_i (what von Neumann called ϵ), and "are being relied upon to behave unreliably in this exact amount" [Pip90] (see also [DO77a, DO77b, Pip89]). Similarly, in our work we assume either full knowledge of p_i, or only knowledge of the first $O(\log T)$ bits of p_i, where T is the computation time of the network. We show these two options to be equivalent. We then continue and expand to a Markovian model of stochasticity: here p_i depends on the neighboring neurons and the recent history of the system. This richer model can be used to describe various natural phenomena.

Types of Randomness

As for the second question, regarding the type of randomness, we consider any type of random behavior that can be modeled by augmenting the underlying deterministic process, either with independent identically distributed binary sequences (IID), or with Markovian sequences. We then abandon the stochastic coin model and substitute it with asynchronicity of update and with various nondeterministic reactions of the neurons themselves. This will be described in Section 9.6, where we discuss stochastically forgetting neurons (each neuron forgets its activation value with some probability; this forgetting is modeled as resetting the activation value to "0") and also the effect of probabilistic changes in the interconnection between neurons; various other types of randomness are proposed.

Amount of Randomness

The next question we consider is the amount of stochasticity allowed in the model. Von Neumann assumed a *constant failure probability* p in the various gates, independent of the size of the network. Furthermore, he allowed *all components* to behave randomly. Thus, larger networks suffer from more unreliability than smaller ones. In contrast, many later models allowed p to decrease with the size of the network (see discussion in [Pip90]), and others assumed the incorporation of fully deterministic/reliable components in critical parts of the network (see, e.g., [MG62, Ort78, Uli74]). Because our network is recurrent, it is easy to verify that when the p_is are constant and all neurons are unreliable, no function requiring non-constant deterministic time $T(n)$ is computable by the network. We thus focus on networks that include both reliable/deterministic neurons and unreliable/stochastic neurons characterized by fixed p_is. It is beyond the scope of this book, but it is worth noting that some biological modelings consider networks that combine deterministic and stochastic behavior. (An equivalent computational model can be achieved by allowing all neurons to behave randomly, while forcing the error to decrease polynomially with the parameter $T(n)$.)

Defining the Output Response

As for the question of defining an output for the probabilistic process, we adopt the bounded error approach (as in the definition of BPP, Section 1.5). Given $\epsilon < \frac{1}{2}$, we only consider networks that yield a wrong output on at most a fraction ϵ of the possible computations.

9.1.1   The Model

The underlying deterministic network is as introduced in Chapter 2. The following definition endows the network with stochasticity.

Definition 9.1.1 A stochastic network has additional input lines, called *stochastic lines*, that carry independent identically distributed (IID) binary sequences, one bit per line at each tick of the clock. The distributions may be different on the different lines. That is, for all time $t \geq 0$, the stochastic line l_i has the value 1 with probability p_i ($0 \leq p_i \leq 1$), and 0 otherwise.

Equivalently, stochastic networks can be viewed as networks composed of two types of components: analog deterministic neurons and binary probabilistic gates/coins. A probabilistic gate is a binary gate that outputs "1" with probability $p \in [0, 1]$.

Yet another way to view the same model is to consider it a network of neurons, some of which function deterministically, while others have "well-described faults"; that is, their faulty behavior can be described by a neural circuit. More on this and on other equivalent models appears in Section 9.6.

We define the recognition of a language by a stochastic network using the bounded error model described in Section 1.5. We assume that the number of steps in all computations on an input ω is exactly the same. An equivalent definition is that for any given run on an input ω, if ω is not accepted in time $T(|\omega|)$, then it is rejected. The classification of an input ω in a computation run is the reject or accept decision at the end of that computation. The final decision of ω considers the fraction of reject and accept classifications of the various computations.

Definition 9.1.2 Let $T : \mathbb{N} \rightarrow \mathbb{N}$ be a total function on natural numbers. We say that the language $L \subseteq \{0,1\}^+$ is ϵ-*recognized in time* T by a stochastic network $\mathcal{N}$ if every $\omega \in \{0,1\}^+$ is classified in time $T(|\omega|)$ by every computation path of $\mathcal{N}$ on ω, and the error probability in deciding ω relative to the language L is bounded: $e_{\mathcal{N}}(\omega) < \epsilon < \frac{1}{2}$.

A relatively standard lemma shows that for probabilistic models of computation, the error probability can be reduced to any desired value [Par94]. This indicates that the following complexity class is well-defined.

Definition 9.1.3 S-NET is the class of languages that are ϵ-recognized by stochastic networks for any $\epsilon < \frac{1}{2}$.

9.2 The Main Results

Now that we have defined the model, we are ready to state the theorems about stochastic networks and compare them with deterministic networks. Proofs appear in the following sections.

9.2.1 Integer Networks

In the deterministic case, if the networks are restricted to integer weights, the neurons may assume only binary activations, and the networks become computationally equivalent to finite automata. Similar behavior occurs for stochastic networks.

Theorem 21 *The class* S-NET$_Z$ *of languages that are ϵ-recognized by networks with integer weights is the set of regular languages.*

The case of integer weights is considered only for the sake of completeness. Rational and real stochastic networks are of greater interest.

9.2.2   Rational Networks

In deterministic computation, if the weights are rational numbers, the network is equivalent in power to the Turing machine model. The addition of stochasticity characterized by real parameters increases the power of the rational network beyond the Turing power. The reader is referred to Definition 1.8.3 in Section 1.8 describing prefix nonuniform complexity classes.

Because we focus on the prefix nonuniform class $BPP/\log *$, we first demonstrate that it is a strict subset of $BPP/\log$. Any tally set is in $P/1$, and thus is clearly also in $P/\log$ and $BPP/\log$. On the other hand, consider a tally set S whose characteristic sequence is completely random (say, Kolmogorov random). It cannot be in any class $RECURSIVE/\log *$ because there is not enough information in $O(\log n)$ bits to get the nth bit of S. As a special case of RECURSIVE, choose BPP and conclude that S is not in $BPP/\log *$. (If we remove the prefix requirement, and leave only the strong or full requirement of Ko [Ko87], $BPP/\log *$ does not change, as can be deduced from Hermo [Her96].)

The following theorem states the equivalence between rational stochastic networks and the class $BPP/\log *$:

Theorem 22 *The class* S-NET$_Q$ *of languages ϵ-recognized by rational stochastic networks in polynomial time is equal to $BPP/\log *$.*

Remark 9.2.1 As a special case of this theorem we note that if the probabilities are all rationals, then the resulting polynomial time computational class is constrained to BPP.   □

Recall that BPP is recursive; it is included both in $P/poly$ ([BDG90] pg. 144, cor. 6.3) and in $\Sigma_2 \cap \Pi_2$ ([BDG90] pg 172, Theorem 8.6). It is still unknown whether the inclusion $P \subseteq BPP$ is strict or not.

9.2.3   Real Networks

Deterministic real networks compute the class $P/poly$ in polynomial time. The addition of stochasticity does not yield a further increase in the computational power.

Theorem 23 *Denote by* NET$_R$ $(T(n))$ *the class of languages recognized by real networks in time $T(n)$, and by* S-NET$_R$ $(T(n))$ *the class of languages ϵ-recognized by real stochastic networks in time $T(n)$. Then*

$$\text{NET}_R\left(T(n)\right) \quad \subseteq \quad \text{S-NET}_R\left(T(n)\right)$$
$$\text{S-NET}_R\left(T(n)\right) \quad \subseteq \quad \text{NET}_R\left(n^2 + nT(n)\right).$$

The results for polynomial stochastic networks are summarized in table 9.1.

Weights	Deterministic	Stochastic
$\mathbb{Z}$	regular	regular
$\mathbb{Q}$	P	BPP/log $*$
$\mathbb{R}$	P/poly	P/poly

Table 9.1: The computational power of recurrent neural networks.

9.3   Integer Stochastic Networks

In this section we prove Theorem 21, which states the correspondence between probabilistic automata and integer stochastic networks. The classical definition of a probabilistic automaton (see [Paz71] for example) is as follows:

Definition 9.3.1 We define a probabilistic automaton $\mathcal{A}$ to be the 5-tuple $\mathcal{A} = (Q, \Sigma, p, q_0, F)$ where Q is a finite set of states, Σ is the finite input alphabet, $q_0 \in Q$ is the initial state, and $F \subseteq Q$ are the accepting states. The probabilistic transition function $p(m, q, a)$, where $m, q \in Q$, and $a \in \Sigma$ specifies the probability of getting into state m from a state q and the input symbol a. This transition function satisfies that for all q and a,

$$\sum_{m \in Q} p(m, q, a) = 1 .$$

We consider the double-transition probabilistic automaton which is a special case of probabilistic automata: the transition function may transfer each state-input pair into exactly two states. That is, for all $q \in Q$ and $a \subset \Sigma$, there are exactly two states m_{1qa}, m_{2qa} such that $p(m_{1qa}, q, a), p(m_{2qa}, q, a) > 0$ and $p(m_{1qa}, q, a) + p(m_{2qa}, q, a) = 1$. It is easy to verify that this automaton is indeed equivalent to the general probabilistic automaton (up to a reasonable slowdown in the computation). The proof is left to the reader.

A double-transition probabilistic automaton can be viewed as a finite automaton with $|Q||\Sigma|$ additional input lines of IID binary sequences; these lines imply the next choice of the transition. From the equivalence of deterministic integer networks and finite automata, we conclude the equivalence between probabilistic automata and integer stochastic networks.

Rabin showed that bounded error probabilistic automata are computationally equivalent to deterministic ones ([Paz71], p. 160). Thus, stochastic networks with integer weights are computationally equivalent to bounded error probabilistic automata, and s-NET$_Z$ is the class of regular languages.

9.4   Rational Stochastic Networks

This section is devoted to the proof of Theorem 22, which places rational stochastic networks in the hierarchy of super Turing computation.

We use a generalization of the classical probabilistic Turing machine [BDG90] that substitutes the random coin of probability $\frac{1}{2}$ by a finite set of real probabilities. (A set of probabilities is required in order to describe neurons with different stochasticity.)

Definition 9.4.1 Let $S = \{p_1, p_2, \ldots, p_s\}$ be a finite set of probabilities. A *Probabilistic Turing machine over the set S* is a nondeterministic machine that computes as follows:

1. Every step of the computation can have two outcomes, one chosen with probability p and the other with probability $1 - p$.

2. All computations on the same input require the same number of steps.

3. Every computation ends with *reject* or *accept*.

We denote by BP$[S, T]$ the class of languages recognized in time T by bounded error probabilistic Turing machines with probabilities over S. We use the shorthand notation BPP$[S] = $ BP$[S, poly]$. Similarly, we denote s-NET$_Q$ $[S, T]$ as the class of languages recognized by rational stochastic networks in time T, with probabilities from S.

Lemma 9.4.2 *Let S be a finite set of probabilities, then BP$[S, T]$ and s-NET$_Q$ $[S, T]$ are polynomially time related.*

Proof.

1. BP$[S, T] \subseteq$ s-NET$_Q$ $[S, O(T)]$:
   Let $\mathcal{M}$ be a probabilistic Turing machine over the set S that computes in time T. We simulate $\mathcal{M}$ by a rational stochastic network $\mathcal{N}$ having stochastic streams l_i with probabilities $p_i \in S$. Consider the program:

   > **Repeat**
   >   **If** $(l_i{=}0)$ **then** NextStep(0, cur-state, cur-tapes)
   >   **else** NextStep(1, cur-state, cur-tapes)
   > **Until** (final state)

where NextStep is a procedure that given the current state of a Turing machine, the current tapes, and which of the two random choices to take, changes deterministically to the next configuration of the machine. Using techniques of the type introduced in Chapter 3 this program can be compiled into a network that computes the same function, having no more than a linear slowdown [Sie96b].

2. S-NET$_Q$ $[S, T] \subseteq$ BP$[S,$poly$(T)]$:
It is easy to verify that if a rational stochastic network $\mathcal{N}$ has s IID input channels, then it can be simulated by a probabilistic Turing machine over the same s probabilities.

∎

Next, we differentiate the case in which all probabilities of the set S are rational numbers from the case where S contains at least one real element.

9.4.1 Rational Set of Choices

Consider probabilistic Turing machines with probabilities over the set S, where S consists of rational probabilities only. Zachos showed that in the error bounded model, if the transition function decides its next state uniformly over k choices (k is finite but can be larger than 2), this model is polynomially equivalent to the classical probabilistic Turing machine with $k = 2$ [Zac82]. When the probabilities are rationals, we can substitute them all by a common divisor which is written as $\frac{1}{k'}$ for an integer k'. This process increases the number of uniform choices, and implies polynomial equivalence between probabilistic Turing machines with one fair coin, and probabilistic machines over a set S. We conclude Remark 9.2.1 stating the computational equality between the class S-NET$_Q$ [poly] and the class BPP. Thus, rational stochasticity adds power to deterministic rational networks if and only if the class BPP is strictly stronger than P. Note that S-NET$_Q$ [poly] must be computationally strictly included in NET$_R$ [poly], because BPP is included in P/poly ([BDG90] pg. 144, cor. 6.3).

9.4.2 Real Set of Choices

Lemma 9.4.2 relates probabilistic Turing machines to stochastic neural networks. The lemma below completes the proof of Theorem 22 by showing the equivalence between real probabilities and log prefix advice in the probabilistic Turing model. We define BP$_R(T) = \cup_{p \in [0,1]}BPP[\{p\}, T]$ as the class of languages recognized by probabilistic bounded error Turing machines that use coins of real probability and compute in time T. BP$_Q(T)/\log *$ is similarly defined for rational probabilities, and with the addition of prefix advice. Note that $\mathrm{BP}_Q[\mathrm{poly}]/\log * = \mathrm{BPP}/\log *$.

Lemma 9.4.3 *The classes $\mathrm{BP}_R(T)$ and $\mathrm{BP}_Q(T)/\log *$ are polynomially related.*

Proof.

1. $\mathrm{BP}_R(T) \subseteq \mathrm{BP}_Q(O(T \log T))/\log *$:
Let $\mathcal{M}$ be a probabilistic Turing machine over the probability $p \in [0,1]$ that ϵ-recognizes the language L in time $T(n)$. We show a probabilistic Turing machine $\mathcal{M}'$ having a fair coin, which, upon receiving prefix advice of length $\log(T(n))$, ϵ'-recognizes L in time $O(T(n) \log(T(n)))$. In *italics* we describe the algorithm for the simulation and then bound its error probability.

> *Let p' be the rational number that is obtained from the $\log(T(n))$ most significant bits of the binary expansion of p. The advice of $\mathcal{M}'$ consists of the bits of p' starting from the most significant bits. One coin flip by $\mathcal{M}$ can be simulated by a binary conjecture of $\mathcal{M}'$, which is based on $\log(T(n))$ coin flips of its fair coin. $\mathcal{M}'$ tosses $\log(T(n))$ times and compares the resulting guessed string with the advice to make a binary conjecture. If the guessed string precedes the advice in the lexicographic order, $\mathcal{M}'$ conjectures "0", otherwise $\mathcal{M}'$ conjectures "1".*

The error probability of $\mathcal{M}'$ is the probability that it generates a sequence of conjectures that would yield a wrong decision. Denote by $r = r_1 r_2 \ldots r_T, r_i \in \{0,1\}$ a sequence of T binary bits. Let

$$\mathrm{Pr}_p(r) = p^{\sum_{k=1}^{T} r_k} (1-p)^{T - \sum_{k=1}^{T} r_k}$$

be the probability that r is the sequence of random choices generated by the coin of $\mathcal{M}$, and let

$$\mathrm{Pr}_{p'}(r) = p'^{\sum_{k=1}^{T} r_k} (1-p')^{T - \sum_{k=1}^{T} r_k}$$

be the probability that $\mathcal{M}'$ generates this sequence of binary choices during its computation.

$\forall r$, if $p' < p$, we denote $q = 1 - p$, and $q' = 1 - p'$, so that $q' > q$, and calculate a connection between the two probabilities:

$$\mathrm{Pr}_{p'}(r) = \frac{\mathrm{Pr}_{p'}(r)}{\mathrm{Pr}_p(r)} \mathrm{Pr}_p(r) \leq (\frac{q'}{q})^T \mathrm{Pr}_p(r) \tag{9.1}$$

We next substitute in Equation (9.1)

$$\left(\frac{q'}{q}\right)^T = (1 + \frac{q' - q}{q})^T$$

and use the approximation of small x

$$1 + x \approx e^x \tag{9.2}$$

to obtain the formula:

$$\Pr_{p'}(r) \leq e^{\frac{q'-q}{q}T} \Pr_p(r) \tag{9.3}$$

Denote by B the set of $T(n)$-long binary sequences which are "bad," i.e. misleading conjectures. The error probability of $\mathcal{M}'$ is $\Pr_{p'}(r \in B)$, which is estimated as

$$\Pr_{p'}(r \in B) \leq e^{\frac{q'-q}{q}T} \Pr_p(r \in B).$$

If p' approximates p with first $\log T$ bits then $q' - q \leq \frac{a}{T}$, and

$$\Pr_{p'}(r \in B) \leq e^{\frac{a}{q}} \Pr_p(r \in B).$$

The error probability of $\mathcal{M}'$ is thus bounded by the constant $\epsilon' = e^{\frac{a}{q}}\epsilon$.

2. $\mathrm{BP}_Q(T)/\log * \subseteq \mathrm{BP}_R(O(T^2))$:

Given a probabilistic Turing machine $\mathcal{M}$, having a fair coin and a logarithmically long prefix advice A that ϵ-recognizes a language L in time $T(n)$, we describe a probabilistic Turing machine $\mathcal{M}'$ with an associated real probability p that ϵ'-recognizes L in time $O(T^2(n))$.

The probability p is constructed as follows. The binary expansion of p starts with ".01", i.e. $\frac{1}{4} \leq p \leq \frac{1}{2}$; the following bits are the advice of $\mathcal{M}$.

$\mathcal{M}'$ computes in two phases:

Phase I — Preprocessing: $\mathcal{M}'$ estimates the advice sequence A of $\mathcal{M}$ by tossing its unfair coin $z = cT^2(n), c \geq 1$ times.

Phase II — Simulating the computation of $\mathcal{M}$: $\mathcal{M}'$ simulates each flip of the fair coin of $\mathcal{M}$ by up to $2T(n)$ tosses using the following algorithm:

(a) Toss the unfair coin twice.

(b) If the results are "01" conjecture "0", if they are "10" conjecture "1".

(c) If the results are either "00" or "11" and (a) was called less than $T(n)$, goto (a).

(d) Here (a) was called $T(n)$ times and the decision was not yet made: conjecture "0".

We first bound the error of the estimated advice. Let $\#1$ be the number of "1"'s found in z flips, and define the estimation

$$\tilde{p} = \frac{\#1}{z},$$

which will be used as an estimation of the advice. The Chebyshev formula states that for any random variable x with expectation μ and variance ν and $\forall \Delta > 0$, $\Pr(|x - \mu|) > \Delta) \leq \frac{\nu}{\Delta^2}$. Here x is the sum of IID random variables. The expectation of such z independent Bernoulli trials is $\mu = zp$, and the variance is $\nu = zp(1 - p)$. We conclude that for all $\Delta > 0$,

$$\Pr(|z\tilde{p} - zp| > \Delta) \leq \frac{zp(1 - p)}{\Delta^2} .$$

Because $p(1 - p) \leq \frac{1}{4}$, by choosing $\Delta = \sqrt{cz}$, for a constant c ($c \geq 1$) we get

$$\Pr\left(|\tilde{p} - p| > \sqrt{\frac{c}{z}}\right) \leq \frac{1}{4c} .$$

Thus, if in the first phase $\mathcal{M}'$ tosses its coin $z = cT^2(n)$ times, then the advice is reconstructed with logarithmically many bits and with an error probability ϵ_1 bounded by $\frac{1}{4c}$ (the first two bits "01" are omitted from the guessed advice).

We next prove the correctness of phase II, which is based on Neumann's technique [vN51]. We compute the probability of $\mathcal{M}'$ to guess a bad sequence of coin flips. As above, we denote by r the sequence of binary conjectures of length $T(n)$ generated by $\mathcal{M}'$ during the algorithm, and by B the set of misleading guesses. As the error of $\mathcal{M}$ is bounded by ϵ, and $\mathcal{M}$ uses a fair coin, the cardinality of B is bounded by $2^{T(n)}\epsilon$. We conclude that

$$\Pr\ (r \in B) \le |B| \max_{b \in B} \Pr\ (r = b) \le 2^{T}\epsilon \max_{b \in B} \Pr\ (r = b). \qquad (9.4)$$

The string with maximum probability is 0^{T}. This probability can be estimated as follows:

- The probability of getting the values "00" or "11" in two successive coin flips is $p' = p^2 + (1 - p)^2$. Thus, the probability of ending a coin flip simulation in step (d) of the algorithm is bounded by $p'' = p'^{T(n)}$. Since $\frac{1}{4} \le p \le \frac{1}{2}$, we conclude that $p' \le \frac{5}{8}$ and $p'' \le \frac{5}{8}^{T(n)}$.

- The probability of ending one coin flip simulation with the conjecture "0" is: $\frac{1}{2}(1 - p'') + p'' = \frac{1}{2} + \frac{p''}{2}$.

We thus conclude

$$\max_{b} \Pr\ (r = b) \le \ \Pr\ (r = 0^{T(n)}) = (\frac{1}{2} + \frac{p''}{2})^{T(n)} \qquad (9.5)$$

and we can substitute Equation (9.5) in Equation (9.4) to get:

$$\Pr\ (r \in B) \le 2^{T(n)}\epsilon(\frac{1}{2} + \frac{p''}{2})^{T(n)} \le \epsilon(1 + p'')^{T(n)} \approx \epsilon e^{T(n)\frac{5}{8}^{T(n)}} \qquad (9.6)$$

which is bounded by a small error ϵ_2 for small enough ϵ. The error probability ϵ' of $\mathcal{M}'$ is bounded by

$$\Pr(\text{"wrong advice sequence"}) + \Pr(\text{"bad guess sequence"}) \le \epsilon_1 + \epsilon_2$$

which is also bounded by $\frac{1}{2}$. ∎

Remark 9.4.4 So far we have discussed stochastic networks, defined by adding choices to deterministic networks. We can similarly define the stochastic nondeterministic network by adding choices to nondeterministic networks. When weights are rationals, the latter class is similar to the framework of interactive proof systems ([BDG90] volume I, chapter 11).

9.5 Real Stochastic Networks

In this section we prove Theorem 23 and show that stochasticity does not add power to real deterministic networks. It is trivial to show that stochasticity does not decrease the power of the model; we thus focus on the other direction and prove that S-NET$_R$ [poly] $\subseteq$ NET$_R$ [poly].

We prove this inclusion in two steps. Given a real stochastic network that ϵ-recognizes a language L, the first step describes a nonuniform family $\mathcal{F}$ of deterministic feedforward networks that recognizes L; this creates only a constant slowdown in the computation. The second step describes a deterministic recurrent network that simulates the family $\mathcal{F}$, with a polynomial slowdown of $n^2 + nT(n)$. (We could skip the first step with a more elaborate counting argument in the second step, but we prefer this method for its simplicity of representation.)

Lemma 9.5.1 *Step 1: Let $0 < \epsilon < \frac{1}{2}$, and let L be a language that is ϵ-recognized by a real stochastic network $\mathcal{N}$ of size N in time T. Then, L can also be recognized by a family of nonuniform feedforward real networks $\mathcal{F} = \{\mathcal{N}_n\}_{i=1}^{\infty}$ of depth $T(n) + 1$ and size $cNT(n) + 1$, where*

$$c = \lceil \frac{8\epsilon \ln 2}{(1 - 2\epsilon)^2} \rceil .$$

Proof.

The technique used in this proof is similar to the one used in the proof that BPP $\subseteq$ P/poly [BDG90].

Let $\mathcal{N}$ be a real stochastic network that ϵ-recognizes a language L as above. We show the existence of a family of deterministic networks that recognizes L. Let r be the number of probabilistic gates g_k, $1 \leq k \leq r$; each outputs "1" with probability p_k. For a given input of length n, by unfolding the network to $T(n)$ layers, each a copy of $\mathcal{N}$, we get a feedforward network with $r'_n = T(n)r$ probabilistic gates. Denote this feedforward stochastic network by $\mathcal{N}'_n$.

We pick a string

$$\rho^{i,n} = \rho_1^i \rho_2^i, \ldots, \rho_{r'_n}^i \in \{0, 1\}^{r'_n}$$

at random, with probability $p_{(j \bmod r)}$ that ρ_j^i is "1". Let $\mathcal{N}'_n\{\rho^{i,n}\}$ be a deterministic feedforward network similar to $\mathcal{N}'_n$, but with the string $\rho^{i,n}$ substituting the probabilistic gates (i.e., ρ_j^i substitutes $g_{j \bmod r}$ in level $(j \ \mathrm{div} \ r)$. We now pick cn such strings

$$\rho[n] = (\rho^{1,n}, \rho^{2,n}, \ldots, \rho^{cn,n})$$

at random. The feedforward net $\mathcal{N}_n$ consists of the cn subnetworks $\mathcal{N}'_n\{\rho^{i,n}\}$ ($i = 1 \ldots cn$) and one "majority gate" in the final level. The majority gate takes the output of the cn subnetworks as its input. That is, for each n, the network $\mathcal{N}_n$ computes the majority over cn random runs of the stochastic net $\mathcal{N}$.

We compute the probability that $\mathcal{N}_n$ outputs incorrectly on an input ω of length n. By the definition of $\mathcal{N}_n$, each $\mathcal{N}'_n\{\rho^{i,n}\}$ has the probability ϵ of being wrong. Thus, picking $\rho[n]$ at random, the probability of $\mathcal{N}_n$ to fail is bounded by $B(\frac{cn}{2}, cn, \epsilon)$; this is the probability of being wrong in at least $cn/2$ out of cn independent Bernoulli trials, each having the failure probability ϵ.

We use the bound $B(\frac{cn}{2}, cn, \epsilon) \le (4\epsilon(1 - \epsilon))^{cn/2}$ from [Par94], and choose $c = \lceil \frac{8\epsilon \ln 2}{(1-2\epsilon)^2} \rceil$ to get

$$B(\frac{cn}{2}, cn, \epsilon) < 2^{-n} .$$

This is a bound on the error probability for any individual input of length n. Thus, the sum of failures for all the inputs of length n is less than 1. Hence, there must be at least one choice of random strings $\rho[n]$ that makes $\mathcal{N}_n$ correctly recognize any input of length n. ∎

The above lemma of the 2-step algorithm introduces a family $\mathcal{F}$ of deterministic feedforward networks that decides L. This specially structured family will be shown in the next step to be included in P/poly. More specifically, in the following lemma we complete the proof of Theorem 23 with a construction of a real deterministic network that simulates $\mathcal{F}$.

Lemma 9.5.2 *Step 2: Any language that is recognized by the real family* $\mathcal{F} = \{\mathcal{N}_n\}_{n=1}^\infty$ *described above can also be recognized by a real deterministic recurrent network* $\mathcal{N}_r$. *Furthermore, an input of length n that is recognized by the network $\mathcal{N}_n$ having depth $T(n)$, can be recognized by $\mathcal{N}_r$ in time $O(n^2 + nT(n))$.*

Proof. We remind the reader that the whole family $\mathcal{F}$ was constructed from a single recurrent neural network; call it $\mathcal{N}_2$. Each member $\mathcal{N}_n$ of $\mathcal{F}$ can thus be described by the tuple

$$(\mathcal{N}_2, n, \rho[n])$$

where $\mathcal{N}_2$ is the underlying deterministic recurrent network, n is the index of the network, and $\rho[n] \in \{0, 1\}^{r'_n cn}$.

Let $\tilde{\mathcal{N}}_2$ be any binary encoding of $\mathcal{N}_2$ and $\tilde{\rho}$ be the infinite string

$$\tilde{\rho} = \rho[1] \; 2 \; \rho[2] \; 2 \; \rho[3] \; 2 \cdots .$$

Let $\alpha = \alpha_1\alpha_2\cdots \in \{0, 1, 2\}^\#$ and denote by $\alpha|_6$ the value $\sum_{i=1}^{|\alpha|} \frac{2\alpha_i+1}{6^i}$. This encoding is Cantor-like, and a network can read weights of this form letter by letter in $O(1)$ time each (see Chapter 3). We next construct the recurrent network $\mathcal{N}_r$ that has the weights $\tilde{\mathcal{N}}_2|_6$ and $\tilde{\rho}|_6$ and recognizes the language L. $\mathcal{N}_r$ operates as follows:

1. $\mathcal{N}_r$ reads the input ω and measures its length.

2. $\mathcal{N}_r$ retrieves the encoding $\rho[n]$ from the constant $\tilde{\rho}|_6$. (This takes $\sum_{j \leq n} |\rho[j]| \leq O(n^2)$ as proven in Chapter 4.)

3. $\mathcal{N}_r$ executes the code:

```
Func Net (ω, ρ[n], n, Ñ2);
Var        Yes, No, z: Counter,
           A: Boolean,
           ρ: Real;
Begin
     z=1 ;
     Repeat
          ρ ← retrieve(ρ̃, i, n)        % retrieving ρ^{i,n}
          A ← simulate (Ñ₂, ρ, ω)
          If A = 1 then Increment(Yes) else Increment(No)
          Increment(z)
     Until (z > cn)
     Net ← Return(Yes > No)
End
```

We know that the command "simulate" is feasible from the constructive Turing machine simulation in Chapter 3. Furthermore, it was shown in [Sie96b] how to construct a net from this type of high-level language. This program, as well as the associated network, takes $nT(n)$ steps. Thus $\mathcal{N}_r$ fulfills the requirements of Lemma 9.5.2, and Theorem 23 is proven.   ∎

9.6   Unreliable Networks

In this section we provide a different formulation of stochastic networks. Our von-Neumann like modeling captures many types of random behavior in networks. It can describe the probabilistic "forgetting" neuron:

$$x_i^+ = \begin{cases} \text{regular update} & \text{with probability } 1 - p_i \\ 0 & \text{with probability } p_i \ ; \end{cases} \tag{9.7}$$

as well as the probabilistic "persistent" (asynchronous) neuron:

$$x_i^+ = \begin{cases} \text{regular update} & \text{with probability } 1 - p_i \\ x_i & \text{with probability } p_i \ ; \end{cases} \tag{9.8}$$

and the probabilistic "weakly connected" neuron x_i, which is defined by the update equation:

$$x_i(t+1) = \sigma \left(\sum_{j=1}^{N} \tilde{a}_{ij} x_j(t) + \sum_{j=1}^{M} b_{ij} u_j(t) + c_i \right) ,$$

where

$$\tilde{a}_{ij}^k = \begin{cases} a_{ij}^k & \text{with probability } 1 - p_{ij} \\ 0 & \text{with probability } p_{ij} . \end{cases} \tag{9.9}$$

We generally allow not only for these three types of errors but for more general faults. Each stochastic neuron computes one of d functions which may be more complicated than the above examples.

Definition 9.6.1 A neuron g is said to have a *well-described fault* if it computes d different functions f_i each with a probability $p_{gi} (\sum_{i=1}^{d} p_{gi} = 1)$ such that all the f_i's are deterministically computable by a net in constant time.

It is not difficult to verify the following:

Lemma 9.6.2 *A stochastic network that consists of both well-described faulty and deterministic neurons can be described by our stochastic modeling.*

The proof is left to the reader.

Remark 9.6.3 Note that if all neurons are stochastic ("catastrophic nets"), the stochasticity can no longer be controlled, and no function requiring non-constant deterministic time T is computable by such a network. Introducing a varying error rate or a nonuniform architecture allows one to overcome the catastrophe. □

Similarly, we can define the "Markovian" model of unreliability: unlike the model of independent erroneous neurons, let us consider devices whose unreliable behavior depends on the last c choices of all devices and the last c global states in the network. This better models biological phenomena such as dying neurons, toxication and the Korsakoff syndrome [Fin94]. Note that this model is not strictly Markovian because transitions do not depend only on the global states, but on the choices as well.

Definition 9.6.4 Let g be a well-described faulty neuron with d choices. Let $x(t) \in [0,1]^N$ be the activation vector in time t, and let the vector

$$F_{c,t} = (x(t-c), \ldots, x(t-1)) \in [0,1]^{cN}$$

represent the activation values of the neurons in the previous c steps. Similarly, let $j(t) \in \{1, 2, \ldots, d\}^N$ be the vector of choices made by all neurons in time t, and let the vector

$$J_{c,t} = (j(t - c), j(t - c + 1), \ldots, j(t - 1)) \in \{1, 2, \ldots, d\}^{cN}$$

represent the choices made by all neurons in the previous c steps. Observe that only $x(t - c)$ in $F_{c,t}$ is necessary: the other elements of $F_{c,t}$ can be computed from $x(t - c)$ given the choices $J_{c,t}$; we use this redundancy for simplicity of presentation.

A *c-Markov network* is a network with some unreliable neurons, for which the probabilities p_{gi} are functions of $J_{c,t}$ and $F_{c,t}$; there exist dN stochastic sub-networks that upon receiving $p_{gi}(t)$ as input, output "1" with this probability.

We state without proof:

Theorem 24 *For any well-described fault and any natural constant $c \geq 1$, c-Markovian networks are computationally equivalent to networks of independent unreliability.*

One final equivalent model that we note is the *asynchronous model*. To characterize the behavior of the asynchronous neural networks, we adapt the classical assumption of asynchronous distributed systems: no two neurons ever update simultaneously. An *asynchronous network* is a network with an additional N-level probabilistic gate, g, where level l_i appears with probability p_i ($\sum_i p_i = 1$). At each time t, only processor $g(t)$ updates, and the output is interpreted probabilistically.

9.7   Nondeterministic Stochastic Networks

While nondeterminism has a single definition in digital computing, it has two possible interpretations in analog models. Weak nondeterminism incorporates guesses of random bits into the computation. The strong nondeterministic model incorporates guesses of real numbers [BSS89].

Definition 9.7.1 A *stochastic architecture* $\mathcal{A}$ is a network in which the probabilities are variables v_i. A *nondeterministic stochastic architecture* is an architecture $\mathcal{A}$ that when given an input string ω, guesses the values of the probabilities $v = v_1, v_2, \ldots, v_N$ and outputs a probabilistic response $\mathcal{A}_v(\omega) \in \{0, 1\}$. As $\mathcal{A}_v$ is a stochastic network, it $\epsilon(v)$ recognizes a language $L_{\mathcal{A}_v}$. The language accepted by the architecture is thus

$$L_{\mathcal{A}} = \{\omega \mid \exists v, \epsilon(v) \in (0, \tfrac{1}{2}) : \mathcal{A}_v(\omega) = 1 \text{ with probability } > 1 - \epsilon(v)\} \,.$$

As in the model of computation over the real numbers, we consider weak and strong nondeterminism. We say that L is accepted by $\mathcal{A}$ using a *strong model of nondeterminism* if $v \in \mathbb{R}^N$, and by a *weak model of nondeterminism* when v is a vector of N (non-periodic) rationals represented with $O(T)$ bits.

Lemma 9.7.2 *Let* nondet-S-NET$_U$ [poly] *be the nondeterministic counterpart of* S-NET$_U$ [poly], *where* $U \in \{Q, R\}$, *then* S-NET$_U$ [poly] $=$ *nondet*-S-NET$_U$ [poly]$_{\text{weak}}$ $=$ *nondet*-S-NET$_U$ [poly]$_{\text{strong}}$.

When considering rational weights, this says no more than that the class BPP$/\log *$ is equivalent to its nondeterministic version; it is a simple corollary of the observation that only the first $O(\log T(n))$ bits of the probabilities v are significant in a probabilistic computation.

Chapter 10

Generalized Processor Networks

Up to this point we have analyzed in detail the computational properties of the analog recurrent neural network. From here on we turn to consider more general models of analog computation, and place our network within this wider framework.

In this chapter we prove that the basic homogeneous network is as powerful as a whole class of generalized processor networks. We prove that if one allows multiplications in addition to linear operations in each neuron (*high-order* neural networks), then the computational power does not increase. Furthermore, no increase in computational power (up to polynomial time) can be achieved by replacing the saturated-linear function with any other function that satisfies only basic continuity requirements. Even using different activation functions in different processors does not increase computational power. This result places an upper bound on the computational power of a large class of analog networks, complementary to the lower bounds in chapters 7 and 8.

The class of generalized analog processor networks, of which our network is a particular example, operates in a continuous phase space. A traditional objection to the use of a continuous phase space in computation is that the resulting process seems very sensitive to noise. In digital machines, the basic components are bit registers; noise in this framework may cause a single bit to take on an incorrect binary value. This can be corrected by performing the same computational step multiple times and taking the majority decision for each bit. This can be implemented either by having the machine repeat each step of the computation multiple times, or by using multiple copies of the machine. Such a process makes a digital computer resistant to small amounts of noise. Taking a majority vote cannot be done with a continuous range of values; averaging multiple noisy values does not completely eliminate noise in the analog domain.

As a manifestation of the "linear precision suffices" property from Chapter 4, we demonstrate that all networks belonging to the class of analog processor networks possess a degree of tolerance to noise and inaccuracies in implementation. Such tolerance to architectural errors is not even well defined in digital computational models.

The P/poly upper bound, and the sufficiency of linear precision are not maintained when relaxing the conditions on the activation functions. In the work by Gavalda and Siegelmann [GS98], it is shown that if our model is augmented with a few threshold neurons, the resulting networks are able to compute arbitrary recursive functions in linear time. These networks, called there *arithmetic networks* require exponential precision when the weights are rationals. (Exponential precision is also required if the activation function is continuous but has the "launching" property that iterates of values near 0 grow exponentially.) They are not subject to any fixed precision bound when the weights are reals, even if the weights are polynomial time computable reals only. Recall from Chapter 5 that for such weights and under the constraint of polynomial time, our networks compute only the class P.

The rest of this chapter consists of four sections. The first introduces the class of generalized analog processor networks. The second section includes the definition of the limited precision version of the above. In Section 10.3 we prove the computational equivalence of this class to our basic neural network, which establishes an upper bound for analog processor networks. We conclude with a short section discussing the robustness of these networks.

10.1   Generalized Networks: Definition

In this section, we consider dynamical systems, which we call *generalized analog processor networks*, whose structure is more general than that of our recurrent neural network. Later, these networks will be shown to be at most as powerful as the previously considered model, providing polynomial speedup at most.

Let N, M, and ℓ be natural numbers. A *generalized analog processor network* is a dynamical system D that consists of N processors $x_1, x_2, \ldots, x_N$, and receives its input $u_1(t), u_2(t), \ldots, u_M(t)$ via M input lines. A subset of the N processors, say $x_{i1}, \ldots, x_{i\ell}$, are used to communicate the output of the system to the environment. In vector form, a generalized analog processor network D updates its processors via the equation

$$x^+ = f(x, u),$$

where x is the current state vector of the network, u is an external input (possibly a vector), and f is a composition of functions:

$$f = \vartheta \circ \pi,$$

where

$$\pi : \quad \mathbb{R}^{N+M} \to \mathbb{R}^N$$

is some vector polynomial in $N + M$ variables with real coefficients, and

$$\vartheta : \quad \mathbb{R}^N \to \mathbb{R}^N$$

is any vector function that has a *bounded range* and is *Lipschitz*. That is, for every $\rho > 0$, there exists a constant C such that for all $x, \tilde{x} \in \text{Domain}(\vartheta)$: if $|x - \tilde{x}| < \rho$ then $|\vartheta(x, u) - \vartheta(\tilde{x}, u)| \le C|x - \tilde{x}|$ for any binary vector u. A similar property holds for $f = \vartheta \circ \pi$.

The formal networks described in Chapter 2 have binary values in both input and output channels. In our definition of generalized analog processor networks, the input is still binary but the output is *soft binary* information. That is, there exist two constants α, β, satisfying $\alpha < \beta$ and called the *decision thresholds*, such that each output neuron of D emits a stream of numbers each of which is either smaller than α or larger than β. We interpret the output of each output neuron y as a binary value:

$$\text{binary}(y) = \begin{cases} 0 & \text{if } y \le \alpha \\ 1 & \text{if } y \ge \beta \,. \end{cases}$$

This more general output convention yields no increase in the computational power, with at most a polynomial speedup.

A neural network is a special case of a generalized analog processor network, in which all coordinates of the function ϑ compute the same piecewise linear function σ, and the polynomial π is first order. Definitions of computation and computation under resource constraints are similar in both frameworks.

10.2   Bounded Precision

Let D be a generalized analog processor network

$$x^+ = \vartheta(\pi(x, u)) \,,$$

as above. Let q be a positive integer. Similarly to the definition in Chapter 4, we say that the *q-truncation* of D is the network with dynamics defined by

$$x^+ = q\text{-Truncation} \left[\vartheta(\pi(x, u)) \right] \,,$$

where "q-Truncation" represents the operation of truncating after q bits. The *q-chop* of D is the network with dynamics defined by

$$x^+ = q\text{-Chop} \left[\vartheta(\pi(x, u)) \right] \equiv q\text{-Truncation} \left[\vartheta(\tilde{\pi}_q(x, u)) \right] \,,$$

where $\tilde{\pi}_q$ is the same polynomial π but with coefficients truncated after q bits.

The subsequent observations ensure that round-off errors due to truncation or chopping are not too large.

Lemma 10.2.1 *Assume that D computes in time T, with decision thresholds α, β. Then, there is a constant c such that the function*

$$q(n) = cT(n)$$

satisfies the following property: for each positive integer n, $q(n)$-Truncation(D) computes the same function as D on inputs of length at most n, with decision thresholds

$$\alpha' = \alpha + \frac{\beta - \alpha}{3} \quad \text{and} \quad \beta' = \beta - \frac{\beta - \alpha}{3} .$$

Proof. Let D be a generalized analog processor network, and let $\tilde{D} = q$-truncation(D), with q still to be decided upon. Let δ be the error due to truncating after q bits, that is, $\delta = c_1 2^{-q}$ for some constant c_1. Finally, let ϵ_t be the largest accumulated error in all of the processors by time t. The following estimates are obtained using the Lipschitz property of f:

$$\epsilon_o = 0$$
$$\epsilon_1 = \delta$$
$$\epsilon_t = \delta \sum_{i=o}^{t-1} C^i = \delta \frac{C^t - 1}{C - 1} ,$$

where C is the Lipschitz constant of f for $\rho = 1$. To handle the deviation in values caused by the truncation, we have to bound the error ϵ_t with a constant γ: choosing $\gamma = \frac{\beta - \alpha}{3}$, this translates to the requirement

$$\delta \le \frac{\gamma(C - 1)}{C^t - 1} \le \tilde{C}^{-t} ,$$

for some constant $\tilde{C}$, which is met when δ is the truncation error corresponding to

$$q(n) = \log_2(\frac{1}{c_1}\tilde{C})T(n);$$

this q is linear in T.   ∎

As a corollary of Lemma 4.2.1, and using an argument exactly as in the proof of Lemma 10.2.1, we conclude:

Lemma 10.2.2 *Lemma 10.2.1 holds for the q-chop network as well.*

This lemma is the "linear precision suffices" version for generalized networks. It will allow us to deduce various results regarding the equivalence of the generalized class to the basic recurrent network, and the tolerance of this class to noise in both input and architecture.

10.3 Equivalence with Neural Networks

Definition 10.3.1 Given a vector function $f = \vartheta \circ \pi$ as above, we say that f is *approximable in time* $A_f(n)$ if there is a Turing machine M that computes $T(n)$-chop(f) in time $A_f(n)$ on each input of length n.

Example 10.3.2 If ϑ is approximable, and π has rational coefficients, then $f = \vartheta \circ \pi$ is approximable (as π is approximable in this case). □

Lemma 10.3.3 *Let $\mathcal{L}(T)$ be the class of languages recognized by generalized analog processor networks in time T, for which f is approximable in time A_f and T is computable in time $M(n)$. Then, $\mathcal{L}(T)$ is included in the class of languages recognized by Turing machines in time $O(M(n) + T(n)A_f(T(n)))$.*

To justify the lemma, assume a generalized analog processor network D satisfying the above assumptions. A Turing machine that approximates D receives an input string ω. As a first step it computes the function $T(|\omega|)$ and estimates the required precision $q(n)$. Then it performs an approximate simulation of D with $q(n)$ precision in each step.

Corollary 10.3.4 *Let D be a network that computes in polynomial time T and let f be approximable in polynomial time. Then the language recognized by D is in P.*

Definition 10.3.5 Given a vector function $f = \vartheta \circ \pi$ as above, we say that f is *nonuniformly $F(n)$-approximable in time* $A_f(n)$, if there is a Turing Machine M that computes $T(n)$-Chop(f) using an advice function in $K[F(n),$ poly$(T(n))]$.

Example 10.3.6 Assume a generalized analog processor network D that computes in time T. A polynomial π with general real coefficients is nonuniformly $T(n)$-computable: for each input of size n, the machine receives the first $O(T(n))$ bits of each coefficient as an advice sequence, and then computes the polynomial. □

From the above results, we conclude as follows:

Theorem 25 *Let D be a generalized analog processor network that computes with a function $f = \vartheta \circ \pi$. Assume that ϑ is nonuniformly $F(n)$-approximable in polynomial time. Then there exists a neural network $\mathcal{N}_D$ that recognizes the same language as D and that does so with at most polynomial time slowdown. Furthermore, if ϑ is $F(n)$-approximable in polynomial time and π involves rational coefficients only, then the weights of $\mathcal{N}_D$ are rational numbers as well.*

In summary, adding flexibility to neural networks does not add power, except for possible polynomial time speedup. This flexibility includes:

- Using high-order instead of first-order neurons.

- Using other activation functions instead of the saturated linear function, including the possibility of using different activation functions in different neurons.

- Allowing the output to be "soft binary" rather than pure binary.

10.4   Robustness

We close this chapter with the important property of robustness, which is another consequence of lemma 10.2.2.

Corollary 10.4.1 *For any time T, there exists ϵ_T such that an error of ϵ_T in the activation values of the neurons would not affect the computation up to time T. That is, generalized analog processor networks are mildly resistant to external noise and are tolerant to implementational inaccuracies in the sense that small enough perturbations in weights or in the sigmoidal activation function do not affect the computation, as long as "soft binary" outputs are considered.*

This robustness is a simple consequence of the Lipschitz property of the activation function. A detailed proof involves defining precisely "perturbations of the activation function"; we omit the details.

Chapter 11

Analog Computation

This chapter summarizes previous work in the field of analog computation, allowing us to view our model from this perspective. It also serves as a good introduction to the next chapter.

In the field of analog computation, any physical system observed by an experimenter in a laboratory and any dynamical behavior in nature is perceived as performing a computational process. Beginning from an initial state (input), a physical system evolves in its state space according to an update equation (the computation process) until it reaches some designated state (the output). Such natural processes can be modeled with dynamical systems by identifying a set of internal variables together with a rule that describes the transformation from state to state. Note the similarity to computational processes modeled by discrete automata (Section 1.2).

The main property that distinguishes analog from digital computational models is the use of a *continuous state space*. Other than this feature, there is no agreed upon formal list of characteristics of analog models of computation. Guided by our view of physical systems as performing computation, we suggest the following features (which together form a desirable framework).

1. Physical dynamics is characterized by the existence of *real constants* that influence the macroscopic behavior of the system. In contrast, in digital computation all constants are in principle accessible to the programmer.

2. Another property of the motion generated by a physical system is *local continuity* in the dynamics. Unlike the flow in digital computation, "locally discontinuous" statements of the following forms are not allowed in the analog setup: "tests for 0" or "if $x > 0$ then compute one thing and if $x < 0$ then continue in another computation path."

3. Although the physical system contains internal continuous values, *discreteness of the output* is dictated by the limited precision of the mea-

surement tools used to probe the continuous phase space. This brings us to a discrete I/O, as in the definition of digital computation.

The combination of real values with finite precision measurement tools implies that one cannot copy the real constants characterizing analog systems. As a result, the ability to construct such analog computers is limited.

Smale was the first to insist on a computational model in which the operations are carried on real values, irrespective of their binary representation [BSS89]. Together with Blum and Shub, he introduced a mathematical model for computation that operates in a single time step on real valued registers and allows for real constants as well. This "BSS" model consists of a recursive algorithm and uses an unbounded number of real memory registers. The algorithm is described as a finite interconnection of four types of nodes: input, output, "computation" (of polynomials), and "branch" (exact threshold operation). The fetching of memory variables can occur either sequentially or by explicitly exchanging the values of registers (i, j); the latter is sometimes referred to as the "fifth type node." When constrained to polynomial time, the computational power of the BSS model lies somewhere between P and PSPACE when the coefficients of the polynomials are constrained to rationals, and between P/poly to PSPACE/poly when the coefficients can take general real values [CG97, BCSS98, MM97]. The exact computational power is unknown*.

The BSS model is considered to be a model of computation over real numbers, rather than a model of analog computation. This is because it lacks the property of local continuity in the dynamics (item #2 above), which is a basic feature in frameworks of analog computing. Our model, on the other hand, fits the particular view of an analog computer as described above.

In this chapter, we present various analog models of computation by stressing their interrelation. Hence, we divide these models into broad categories, with partial overlap: discrete time models (Section 11.1), continuous time models (Section 11.2), hybrid models (Section 11.3), and dissipative computational models (Section 11.4).

11.1 Discrete Time Models

We begin our survey with the cellular automaton (CA). This is a computational model that is an infinite lattice of discrete variables with a local homogeneous transition rule. CAs contain Turing machines as a special case, and are thus computationally universal [SI71]. When the variables are reals, the machine is sometimes called an analog cellular automaton, or a coupled

---

*A variant, the weak BSS model [Koi93], is computationally equivalent to ARNN.

map lattice (CML) [OM96]. It can be thought of as a generalization of both the BSS model and our networks. The BSS model is equivalent to a Turing machine with real valued cells, and hence is easily simulated by a CML. Our neural network can be simulated by a CML, where every cell computes as a neuron and the neighborhood structure is large enough to include the finite size of the simulated network. As a result, even a finite CML is computationally stronger than a universal Turing machine. Cellular automata and coupled map lattices are used in modeling a broad class of physical phenomena [TM87].

Coupled map lattices can be considered time and space discretizations of PDEs, which are thus computationally universal as well [Omo84]. This leads us to another group of analog computational models, ones that update in continuous time.

11.2 Continuous Time Models

The line of work regarded as the "general purpose analog computer" (GPAC) was dominated by Shannon [Sha41], Pour-el [Pou74, PR88], and Rubel [Rub81, Rub89, Rub93]. It describes a mathematical abstraction that consists of finite numbers of boxes with plenty of feedback. The boxes are of five types: one produces the running time, one produces any real constant, one is an adder, one is a multiplier and the last and crucial one is an integrator. Although the GPAC was originally suggested as a general computational machine to handle computational physics, it was found to be too weak to solve even some very standard problems, such as the Dirichlet problem, or being able to generate the Euler gamma function. The "general purpose analog computer" produces only solutions of initial-value problems for algebraic ordinary differential equations [Rub93]. The important questions raised by these seminal works inspired many researchers to focus on continuous differential equations as a computational device.

One fundamental question is which functions can be computed with differential equations. ODEs were used to simulate various discrete time models, thus providing lower bounds on their computational power. Brockett demonstrated how to simulate finite automata by ODEs [Bro93]; Branicky generalized this result to simulate Turing machines [Bra94], thus proving their computational universality. These systems retain the discrete nature of the simulated system, in that they follow the computation of a given discrete map step by step via a continuous time equation. Upper bounds on the power of different types of ODEs are yet to be found. Two new articles [SF98, SBF98] develop a theory facilitating complexity analysis of computation for ODEs. These papers focus on attractor dynamics and will be described in Section 11.4.

11.3   Hybrid Models

Motivation for some of the work on hybrid models of discrete and continuous time computation is derived from the realization that the functional-
ity of controllers is not fully captured by discrete time dynamics, but it is
also not naturally described by continuous-time dynamics. Hybrid systems
combine discrete and continuous time dynamics, usually by means of ODEs
that are governed by finite automata. Among famous hybrid models are the
works by Tavernini [Tav87], Back-Guckenheimer-Myers [BGM93], Nerode-
Kohn [NK93], the practical stepper motors and transmissions by Brockett
[Bro89, Bro91, Bro93], Branicky [Bra93, Bra95], and stabilizing properties by
Artstein [Art96]; more can be found in [ALS93, AMP95, BC96, GNRe93].
Due to their finite automaton component, hybrid systems do not adhere to
the feature of local continuity.

11.4   Dissipative Models

Dynamical systems are called dissipative if their dynamics converge to attractors. When a dissipative system has an energy (Lyapunov) functional, the
attractors are fixed points [HS74]; otherwise, more complex attractors may
appear.

Dissipative systems are popular in neural modeling of memory. Content
addressable memory allows one to recall a stored word without referring to
its physical location; associative memory allows for recall based on partial or
partially erroneous information. Neural networks with an energy functional
can implement both types of memory. Memories are encoded (or loaded)
in the local minima states by setting the parameters of the system. When
memories are all loaded the network can be used to "complete" or "clean" a
partial/noisy picture, by presenting it as an initial condition and observing
the flow to the appropriate fixed point attractor. The initial state is thus
equivalent to the input of a computational problem, the evolution along the
trajectory is the computation process, and the attractor describes the solution.
The most well-known neural memory model is probably the Hopfield network
[Hop84, HT85, HKP91].   Other related networks are described by Amari,
Anderson, Kohonen, Nakano, Willshaw, Little, Atiya and Abu-Mostafa, and
Morita [Ama71, And72, And68, Koh72, Nak72, NN71, WBL69, Lit74, LS78,
AA93, Mor93]. The meaningful attractors of these networks, where infor-
mation is stored, are all simple: either stable fixed points or limit cycles.
Nevertheless, there are various models of neural activity that report chaos,
but do not consider the resulting chaotic systems in computational terms.

A general theory of computation for dissipative dynamical systems was
developed by Siegelmann, Fishman and Ben-Hur [SF98, SBF98]. Computers

can be described as dynamical systems evolving in a discrete configuration space and updating in discrete time. Physical systems are similarly modeled by dynamical systems; however, these are often ODEs which function in continuous phase space and evolve in continuous time. The above mentioned theory interprets the evolution of dissipative dynamical systems, both discrete and continuous, as a process of computation. This work not only develops a fundamental view of physical systems as being special-purpose analog computers, it is also useful for the complexity analysis of the many ODE based computer algorithms (see e.g., [HM94]). Prior to this theory, ODE based algorithms could only be analyzed by means of time discretization. Now analysis can be carried out directly in the continuous domain.

Attractor systems exhibit various levels of complexity, and their long term dynamics may converge to fixed points or exhibit complex chaotic behavior. The attractor to which a system flows from the initial condition can be viewed as the output. Generally, there is no way to analytically determine the basins of attraction and their boundaries, and thus the output is unpredictable. The time it takes the system to converge to the vicinity of an attractor is measured in terms of the characteristic time scale of the system. This new complexity measure replaces the number-of-steps measure. Dissipative dynamical systems are then classified into the computational complexity classes P_d, Co-RP$_d$, NP$_d$ and EXP$_d$, which are defined relative to the new measure. While fixed points can be computed efficiently, chaotic attractors can only be computed efficiently by means of nondeterminism. The inherent difference between fixed points and chaotic attractors leads us to propose that, in the realm of dynamical systems, efficient deterministic computation differs from efficient nondeterministic computation and the two do not coincide: $P_d \neq$ NP$_d$.

The class LIN$_d$, a subclass of P_d, is defined as the set of problems solvable optimally (in linear time) by dissipative dynamical systems. This class contains, for example, many special cases of linear programming [PS82, Kar91], which is the problem of maximizing a linear function subject to linear inequality and equality constraints. The feasible solutions to a linear programming problem form the vertices of a polyheder. The simplex algorithm that searches directly in the vertices' domain has exponential worst case complexity. In contrast, interior point algorithms and gradient flows (e.g. Karmarkar [Kar84] and the work of Faybusovich [Fay91]) approach the solution from inside the polytope and require polynomial time only. We claim that a flow based on the principle of the Karmarkar algorithm frequently has optimal (linear) complexity. P and P_d are conjectured to be closely related, but the precise relation between the dissipative computational classes and the classical theory has yet to be resolved.

This overview of the field of analog computation provides the background for the next chapter, where we suggest our analog network as a standard for analog super-Turing theories.

Chapter 12

Computation Beyond the Turing Limit

This chapter differs from the rest of the book. In addition to containing mathematical proofs, it also discusses possible philosophical consequences in the realm of super-Turing theories.

Since 1936, the standard and accepted model of universal computation has been the Turing machine, which forms the basis of modern computer science [HU79]. The Church-Turing thesis (C-T), the prevailing paradigm in computer science, states that no possible abstract digital device can have higher capabilities (except for relative speedups due to more complex instruction sets or parallel computation) than Turing machines. This thesis identifies the notion of an algorithm with a computation within the Turing model, and it identifies efficient problem solving with polynomial time Turing computation. Although the C-T thesis focuses mainly on mathematical modeling, it is nonetheless frequently perceived as reflecting the physical limitations of computation in nature. This expanded interpretation of the C-T thesis, also called the physical C-T thesis, states that no realizable computing device can be more powerful than a Turing machine [Deu85].

Over the last few decades, some voices argued that although the Turing model is indeed able to simulate a large class of computations, it does not necessarily provide a complete picture of the computation possible in nature. These voices have been calling for additional theories, algorithms or media on the basis of natural processes which the current digital computer model overlooks. Feynman suggested making use of the non-locality of quantum physics [Fey86]; Penrose claimed that a super-Turing model of the human brain is required in order to capture the phenomenon of biological intelligence [Pen89, Pen94]. Among the novel suggestions are the quantum Turing machine [Sho94, Ben82a, Ben82b, BV93, BB92, Deu85, Deu89, Fey82, Fey86] and the DNA computer [Adl94, Lip95]. Both of these models are conjectured to compute in polynomial time recursive functions which are not in P. However, they cannot compute nonrecursive functions, and in this sense they are not super-Turing models.

In the previous chapter we introduced the not yet fully sculpted general framework of analog computation, and contrasted it with digital computation. The neural network model, which fits in this framework, was proven throughout this book to be stronger than the standard digital model. The two become equivalent only when certain constraints are imposed. The neural network has some degree of robustness, and computationally represents a rich class of analog processor networks. In this chapter we demonstrate that our simple neural network encompasses many other models of analog computation, and thus suggest it as a basic representative of analog systems. From here we can dare propose an analogy to the mathematical Church-Turing thesis of computability. The thesis of time bounded analog computation reads as follows:

> *No possible abstract analog device can have more computational capabilities*
> *(up to polynomial time) than first-order recurrent networks.*

Note that this thesis can also be stated for the stochastic case; this is because P/poly and its stochastic version are computationally equivalent (see Chapter 9).

It is obvious that the extra power of neural networks stems from their non-rational weights. We next suggest an alternative interpretation of this thesis. It can be perceived as differentiating between static computational models and dynamically evolving ones with learning capabilities. The classical computing paradigms are static; they include only rational constants and are bounded by the Turing power. Evolving machines, on the other hand, can tune their internal constants/parameters, possibly on some continuum where the values would not be measurable by an external observer.

We showed that "linear precision suffices," so that parameters need not be fully tuned when beginning the computation. On the contrary, the machine can start with rational parameters, and increases its precision over time. The combined process of learning and computation creates the super-Turing computation. It is reasonable to suggest that the speed and accuracy of the learning process are the two parameters that determine the network's exact location on the computational continuum, ranging from the Turing machine (no learning) all the way up to our neural network (highest level learning). In conclusion, the neural network model is not only a basic representative of analog systems, but it can also be perceived as a parametric model of learning machines.

Up to this point, we have discussed the mathematical perspective of the analog computation thesis; now we are ready to raise issues related to the physical world. We wonder whether there are indeed physically oriented models that cannot be mimicked by the Turing machine, but can be simulated with our super-Turing model. An affirmative response is provided by considering chaotic systems [GOY87]. Due to their sensitivity to minute changes in parameter values, the behavior of chaotic systems can only be approximated by Turing machines up to a certain point in time, after which the level of precision is too low.

Consider for example the Henon map, defined by:

$$x_{n+1} = a + by_n - x_n^2$$
$$y_{n+1} = x_n ,$$

for constants a and b [GOY87]. The behavior of this system is very sensitive to the choice of its constants. For $a = 1.3$ and $b = 0.3$, the system cycles in a 7-period cycle. When the constant a is minutely increased, the system moves into a 14-period cycle, then into a 28-period cycle, etc. For a further small increase in a, the system gets into a chaotic motion; see Figure 12 (which was plotted using *Dynamics* [NY94]). Many other examples of chaotic systems sensitive to parameter changes can be found in [GOY87, BG90, Ott93]. Because of the real constants, the dynamics is defined on a continuous rather than discrete phase space, and cannot be described in the Turing model.

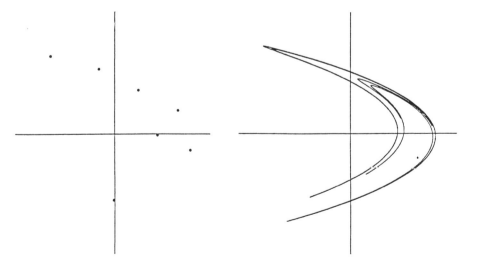

Figure 12.1: The Henon map for different constants

Both chaotic systems and neural networks are defined with, but do not require, full precision. For both, linear precision suffices, suggesting that the analog neural network is indeed a natural representation of chaotic (idealized) physical dynamics.

We are now facing a fundamental question regarding the relevance of P/poly to natural processes. Even if, as argued above, learning machines are super-Turing and neural networks *do* mimic some natural phenomena, we must still verify that our neural network is not too powerful computationally, in order to conclude that there is no need to search for another analog representative between Turing machines and neural networks. This is the focus of the rest of this chapter.

The issue of "not being too strong" is debatable, and must be argued on intuitive grounds. For some, the class P includes all physically efficient computations. Stochasticity might be considered physically relevant, indicating the possible need for a probabilistic computational class such as BPP. Others would insist on the class BPP/ log * that makes moderate use of real parameters. We argue that P/poly is a feasible analog model. For this purpose we define particular dynamical systems, based on the shift map, that may describe idealized physical phenomena. We prove that our networks are computationally equivalent to these maps, and hence constitute a relevant (not too powerful) model. A corollary shows how to obtain a variant of the classical RAM machine that will be equivalent to our model.

This last chapter is organized as follows. In Section 12.1 we describe the analog shift map; we prove in Section 12.2 that it is computationally equivalent to our network. In Section 12.3 we describe the physical plausibility of the analog shift map. We conclude this chapter and the book with a speculative note.

12.1   The Analog Shift Map

In the literature of dynamical systems, chaos is commonly exemplified by the *shift map* (such as the Baker's map [AA68] or the Horseshoe map [GH83]) over a set of bi-infinite dotted sequences, defined below. Here, we present a particular chaotic dynamical system that is similar to the shift map and the *generalized shift map* [Moo90], and computationally is as strong as analog neural networks. We call this dynamical system the *analog shift map*.

Let Σ be a finite alphabet. A dotted sequence over Σ (denoted by $\dot{\Sigma}$) is a sequence of letters of which exactly one is the dot sign '.' and the rest are all in Σ. The dotted sequences can be finite, (one-side) infinite, or bi-infinite over Σ.

Let $k \in \mathbb{N}$ be an integer, the shift map

$$S^k : \dot{\Sigma} \to \dot{\Sigma} : (a)_i \mapsto (a)_{i+k}$$

shifts the dot k number of places, where negative values imply a left shift and positive ones a right shift. For example,

$$S^3(\cdots a_{-2}\, a_{-1}\, .\, a_1\, a_2\, a_3\, a_4\, a_5 \cdots) = \cdots a_{-2}\, a_{-1}\, a_1\, a_2\, a_3\, .\, a_4\, a_5 \cdots\, .$$

The generalized shift map is defined by Moore [Moo90, Moo91] as follows: a finite dotted substring is replaced with another dotted substring according to a function G, then this new sequence is shifted an integer number of places either left or right according to a function F. Formally, the generalized shift is the function

$$\Phi : a \mapsto s^{F(a)}(a \oplus G(a)) , \tag{12.1}$$

where the function $F : \dot{\Sigma} \to \mathbb{Z}$ indicates the amount of shift (where negative values cause a shift left and positive ones a shift right), and the function $G : \dot{\Sigma} \to \dot{\Sigma}$ describes the modification of the sequence. Both F and G have finite *domains of dependence (DoD)*, that is, F and G depend only on a finite dotted substring of the sequence on which they act. G has a finite *domain of effect (DoE)*, i.e. every sequence in the image of G consists of a finite dotted sequence, padded to both sides by infinitely many ϵ's, where ϵ is the "empty element," not contained in Σ. Note that the DoD and DoE of G do not need to have equal length. Finally, the operation $\oplus$ is defined by

$$(a \oplus g)_i = \begin{cases} g_i & \text{if } g_i \in \Sigma \\ a_i & \text{if } g_i = \epsilon \ . \end{cases}$$

The generalized-shift function is homeomorphic to the action of a piecewise differentiable map on a square Cantor set. Moore conjectured that such maps arise in physical systems consisting of a free particle moving between plane mirrors. Most interestingly for the present discussion, Moore proved that the generalized-shift map is computationally equivalent to a Turing machine. This result, thus, connects chaotic dynamical systems with the classical computational model.

Here, we introduce a new chaotic dynamical system: the "analog shift map." It is similar to the generalized shift function in Equation (12.1), except that the substituting dotted sequence (DoE) defined by G is allowed to be finite, infinite, or bi-infinite, rather than finite only. The name "analog shift map" implies a combination of the shift operation with the computational power of analog computational models, as will be proved later.

Example 12.1.1 To illustrate, assume the analog shift defined by

DoD	F	G
0.0	1	$\pi.$
0.1	1	.10
1.0	0	1.0
1.1	1	.0

Here $\bar{\pi}$ denotes the left infinite string "$\cdots 51413$" in base 2 rather than base 10. The table is a short description of the dynamical system, in which the next step depends upon the first letter to the right of the dot and the one to its left. If these letters (i.e. the DoD) are "0.0", then (according to the first row of the table) the left side of the dot is substituted by "$\bar{\pi}$" and the dot moves one place to the right. If the DoD is instead "0.1" (as in the second row of the table), the second letter to the right of the dot becomes "0" and there is a right shift; etc.

The dynamic evolving from

$$000001.10110$$

is as follows: here the DoD is "1.1", hence (by the fourth row of the table) the letter to the right of the dot becomes "0" and the dot is shifted right:

"1.1": (000001.00110) 0000010.011

Now, the DoD is "0.0":

"0.0": ($\bar{\pi}$.0110) $\bar{\pi}$0.110
"0.1": ($\bar{\pi}$0.100) $\bar{\pi}$01.00
"1.0": ($\bar{\pi}$1.00) $\bar{\pi}$01.00

Here the DoD is "1.0" and no changes occur, this is a fixed point.   □

The computation associated with the analog shift systems is the evolution from the initial dotted sequence until reaching a fixed point (or a limit cycle), from which the system does not evolve any further. The computation does not always end; when it does, the input-output map is defined as the transformation from the initial dotted sequence to the final subsequence to the *right of the dot*. (In the above example, a fixed point is reached in four steps, and the computation was from "000001.10110" to "00".) To comply with computational constraints of finite input/output, attention is confined to systems that start with finite dotted sequences and halt with either finite or left infinite dotted sequences only. Even under these finiteness constraints, the analog shifts describe richer maps than Turing machines do; they are computationally equivalent to neural networks.

12.2   Analog Shift and Computation

We next prove the computational equivalence between the analog shift map and the neural network model. Denote by $\text{SHIFT}_A(T)$ the class of functions computed by the analog shift map in time T, and by $\text{NET}_R(T)$ the class of functions computed by a real weights network in time T. The following theorem states the polynomial equivalence of these two classes.

Theorem 26 *Let F be a function so that $F(n) \geq n$. Then*
$\text{SHIFT}_A (F(n)) \subseteq \text{NET}_R (\text{Poly}(F(n)))$, *and*
$\text{NET}_R (F(n)) \subseteq \text{SHIFT}_A (\text{Poly}(F(n)))$.

Proof. We assume, without loss of generality, that the finite alphabet Σ is binary; $\Sigma = \{0, 1\}$.

1. $\text{SHIFT}_A (F(n)) \subseteq \text{NET}_R (\text{Poly}(F(n)))$:
   Given a bi-infinite binary sequence

   $$S = \cdots a_{-3}\, a_{-2}\, a_{-1} \,.\, a_1\, a_2\, a_3 \,\cdots,$$

   we map it into the two infinite sequences

   $$S_r = .\, a_1\, a_2\, a_3 \,\cdots \qquad\qquad S_l = .\, a_{-1}\, a_{-2}\, a_{-3} \,\cdots.$$

   A step of the analog shift map can be redefined in terms of the two infinite sequences S_l and S_r rather than the bi-infinite sequence S itself:

   $$\tilde{\phi}(S_l, S_r) = (S_l \oplus G_l(d_l, d_r),\, S_r \oplus G_r(d_l, d_r)).$$

   Here, $d_l = a_{-1}\, a_{-2} \cdots a_{-d}$ and $d_r = a_1\, a_n \cdots a_d$, assuming, without loss of generality, that the DoD is of length $2d$ and is symmetric; that is, $|d_l| = |d_r| = d$. The DoE of the binary sequences $G_l(d_l, d_r)$ and $G_r(d_l, d_r)$ may be unbounded.

   We next prove the computational inclusion of the analog shift map in neural networks. For this aim, we describe an algorithm that can be compiled into a network by using the techniques developed in Chapters 3 and 4. (This was lately generalized to a high level programming language NIL [Sie96b]). In the following algorithm, we consider the binary sequences S_l and S_r as unbounded binary stacks; we add two other binary stacks of bounded length, T_l and T_r, as temporary storage. We use the stack operations described in Section 3.2: Top(stack), which returns the top element of the stack; Pop(stack), which removes the top element of the stack; and Push(element, stack), which inserts an element on the top of the stack; we do not need the non-empty predicate for our simulation. The computational inclusion is shown in four steps:

   (a) Read the first d elements of both S_l and S_r into T_l and T_r, respectively, and remove them from S_l, S_r.

   > **Procedure** Read;
   > **Begin**
   >   **For** $i = 1$ to d
   >   **Parbegin**

$$T_l = \text{Push (Top } (S_l), T_l), \; S_l = \text{Pop } (S_l);$$
$$T_r = \text{Push (Top } (S_r), T_r), \; S_r = \text{Pop } (S_r);$$

Parend

End;

The same task could be expressed concisely as the sequence $\text{Push}^d(S_l, T_l)$, $\text{Push}^d(S_r, T_r)$, $\text{Pop}^d(S_l)$, $\text{Pop}^d(S_r)$, where the Push and Pop operations are generalized, to be executed d times, for any finite d.

(b) For each choice of the pair

$$\nu_i = (\xi_l^i, \xi_r^i) \in \{0,1\}^d \times \{0,1\}^d \quad i = 1, \ldots, 2^{2d}$$

of the DoD, there is an associated pair of substituting strings:

$$(\mu_l^i, \mu_r^i) \in \{0,1\}^{\kappa_l^i} \times \{0,1\}^{\kappa_r^i}$$

(of the DoE), where each length κ_v^i ($i = 1, \ldots, 2^{2d}$, $v \in \{l, r\}$) is either bounded by some constant k or is ∞. We also consider μ's as stacks.

(c) Computing the $\oplus$ function.

Procedure Substitute(μ_l, μ_r, S_l, S_r);
Begin
    Parbegin
        If$(\kappa_l > k)$ (*μ_l is infinitely long *)
            then $S_l = \mu_l$
            else $S_l = \text{Push }^{\kappa_l}(\mu_l, S_l)$;
        If$(\kappa_r > k)$ (* The parallel case for r *)
            then $S_r = \mu_r$
            else $S_v = \text{Push }^{\kappa_r}(\mu_r, S_r)$;
    Parend
End;

The following program simulates one step of the analog shift map:

Program AS-step();
Begin;
    Read;
    Substitute
End;

That is, there is a network that computes equivalently to the analog shift map. Its exact architecture is irrelevant for our current purpose but it can be constructed by the methods developed in [Sie96b] from the algorithm above.

2. $\text{NET}_R\left(F(n)\right) \subseteq \text{SHIFT}_A\left(\text{ poly }(F(n))\right)$:

We next show how to simulate a Turing machine $\mathcal{M}$ with polynomial advice via an analog shift map. Because $\text{NET}_R\left(\text{poly}\right) = \text{P}/\text{poly}$ this will prove the result. We will use the following observation. Denote the left infinite string that is the concatenation of all advice of $\mathcal{M}$ by

$$\nu = \langle \ldots, \nu(3), \nu(2), \nu(1) \rangle,$$

and the concatenation of the first n advice by

$$\nu'(n) = \langle \nu(n), \ldots, \nu(2), \nu(1) \rangle.$$

Constrained by polynomial computation time and polynomially long advice, it is easy to verify that a Turing Machine that receives the advice $\nu'(n)$ is equivalent to a Turing machine that receives the advice ν. (The machine cannot access more than the first polynomially many advice bits during polynomial time computation.)

We now show how to simulate a Turing machine with polynomial advice via an analog shift map; our simulation is similar to the one made by Moore, but some preprocessing is required. A configuration of the Turing machine consists of its tape, the relative location of the read/write head in the tape, and its internal control state. Moore encoded the configuration in the bi-infinite string using the fields:

| 0^∞ | | tape - left | | . | | state | | tape - right | | 0^∞ |

That is, the string starts with infinitely many "0"'s followed by the non-empty part of the tape to the left of the read/write head, then the binary point, the internal state of the machine, the part of the tape under the head and to its right, and again infinitely many "0"'s that encode the empty part of the tape. In each step of Moore's simulation, the DoD contains the state, the tape letter under the read/write head, and the two letters surrounding it. The DoE is such that the operation may simulate writing a new tape letter, entering a new control state, and moving the head one step to either the right or the left.

In our simulation, we allow for more flexible encoding of the Turing machine configuration:

| garbage | | ## | | tape - left | | . | | state | | tape - right | | 0^∞ |

We substitute the 0^∞ string to the left of the tape with two fields: garbage and "##". Here, garbage means infinitely many bits with no relevant meaning, and "##" marks the left-end of the non-empty part of the tape.

We suggest the following encoding that will allow clear interpretation of the string: "10" will present the "0" letter on the tape, "11" will present the letter "1" on the tape; "01" will present the left-end marker. The internal state of the machine will be encoded by a sequence of "0"'s only and will end with "1"; and 0^∞ still denotes the infinite sequence of "0"'s that represents the empty right part of the tape.

Assume the Turing machine has the set of internal states $\{p_1, \ldots, p_Q\}$. We add the dummy states $\{q_1, \ldots, q_r\}$ for a constant r. Now the Turing Machine with a polynomial advice will act as follows:

(a) The initial bi-infinite string is

$$\boxed{0^\infty} \quad \cdot \quad \boxed{q_1} \quad \boxed{\omega} \quad \boxed{0^\infty}$$

where ω is the encoded input string.

(b) In the next step the string is transferred to

$$\boxed{\nu} \quad \cdot \quad \boxed{q_2} \quad \boxed{\omega} \quad \boxed{0^\infty}$$

where $\nu = \langle \cdots, \nu(3), \nu(2), \nu(1) \rangle$.

(c) In polynomially many steps, the part $\langle \nu(|\omega|-1) \ldots \nu(1) \rangle$ is removed, and the remaining part of the infinite string ν is partitioned into the relevant part $\nu_{|\omega|}$ and the garbage part $\langle \ldots \nu(|\omega|+2)\nu(|\omega|+1) \rangle$.

$$\boxed{\text{garbage}} \quad \boxed{\#\#} \quad \boxed{\nu(|\omega|)} \quad \cdot \quad \boxed{q_r} \quad \boxed{\omega} \quad \boxed{0^\infty}$$

This is done by a recursive algorithm, linear in $|\omega|$, and thus can be executed by a Turing machine, or equivalently, by a GS map. The relevant advice $\nu(|\omega|)$ is next transferred to the right side of the tape:

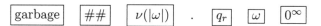

$$\boxed{\text{garbage}} \quad \boxed{\#\#} \quad \cdot \quad \boxed{p_1} \quad \boxed{\omega} \quad \boxed{\nu(|\omega|)} \quad \boxed{0^\infty}$$

where p_1 is the initial state of the advice Turing machine.

(d) From now on, each step of the machine is simulated as by Moore, with the single difference being that the left-end marker should be preserved in the location immediately to the left of the non-empty part of the tape.

∎

Corollary 12.2.1 *Consider the standard bit RAM machine [BDG90]. If we allow it to fill up infinitely many bit registers in one step, the resulting model is equivalent to P/poly. It is also equivalent to a Turing machine that allows for an entire (infinitely long) side of the tape to be written in one step.*

This corollary suggests a way of transforming digital machinery into analog machinery. It stresses the difficulties of implementing analog computers, yet does not reduce the value of the analog model, as it provides a mathematical model expressing physical phenomena, which will be discussed in the next section.

12.3 Physical Relevance

The appeal of the analog shift map is not only as a classical chaotic dynamical system that is associated with the analog computation model. It may also serve as a mathematical description of idealized physical phenomena. The idealization allows for model assumptions, such as the choice of a convenient scale, a noise free environment, and the physics of a continuous medium. Some physical models previously considered equivalent to Turing machines are actually stronger: they are exactly computationally equivalent to analog models. This assertion can be demonstrated, for example, by the system introduced by Moore [Moo90], as a possible realization of the generalized shift map.

The system is a "toy model" describing the motion of a particle in a three dimensional potential, such as a billiard ball or a particle bouncing among parabolic mirrors. A finite number of mirrors suffices to describe the full dynamics: one mirror for each choice of the DoD. The (x, y) coordinates of the particle, when passing through a fixed, imaginary plane $[0, 1] \times [0, 1]$, simulate the dotted sequence "$x.y$". To define the computation, the particle starts in input location $0.y$ where y is the finite input string ω; the output is defined in finite terms as well.

We consider the same mirror system, but we see it as an analog system. The main difference between Moore's view and ours is in the amount of information required to describe the angles of the mirrors. Moore insists on describing the angles of the mirrors with a fixed number of bits. For us, the characterizations of the mirrors cannot be fully described finitely, although we are not necessarily interested in anything more than a certain degree of precision in any particular computation. This infiniteness in the description of the mirrors is modeled by the infinite DoE's of the function G (only a finite number of mirrors is required because the DoD is finite). We can make the end of computation observable by forcing the input and output to reside in observable areas by using, for example, Cantor set encoding. Another possible system may be based on the recent optical realization of Baker's map [KHD94].

These different views of the same system correspond to the difference between the generalized shift map and the analog shift map, or between Turing machines and analog models of computation. Our goal is more ambitious than Moore's; while he takes from physical systems only the fixed number of bits that will allow him to perform a process of computation in the Turing paradigm, we insist on considering the full amount of information provided by physical systems, and move from the classical Turing model to one which will describe these systems.

12.4   Conclusions

The basic neural network model described in this book encompasses many super-Turing analog systems. It is also computationally equivalent to a certain idealized chaotic physical system. Based on these equivalences, we propose that, in the realm of analog computation, our neural network be considered a standard model, functioning in a role parallel to that of the Turing machine in the Church-Turing thesis. As such, it becomes a point of departure for the development of alternative computational theories.

Bibliography

[AA68]   V. I. Arnold and A. Avez. *Ergodic Problems of Classical Mechanics.* Benjamin, New York, 1968.

[AA93]   A. F. Atiya and Y. S. Abu-Mostafa. An analog feedback associative memory. *IEEE Transactions on Neural Networks*, 4:117–126, 1993.

[Adl78]   L. Adleman. Two theorems on random polynomial time. In *Proc. Nineteenth IEEE Symposium on Foundations of Computer Science*, pages 75–83, New York, 1978. IEEE Computer Society.

[Adl94]   L. M. Adleman. Molecular computation of solutions to combinatorial problems. *Science*, 266(5187):1021–1024, November 11 1994.

[ADO91]   N. Alon, A. K. Dewdney, and T. J. Ott. Efficient simulation of finite automata by neural nets. *Journal of the Association for Computing Machinery*, 38(2):495–514, April 1991.

[ALS93]   P. J. Antsaklis, M. D. Lemmon, and J. A. Stiver. Hybrid systems modeling and event identification. Technical Report ISIS-93-002, University of Notre Dame, Indiana, 1993.

[Ama71]   S-I Amari. Characteristics of randomly connected threshold-element networks and network systems. *Proceedings of the IEEE*, 59:33–47, 1971.

[AMP95]   A. Asarin, O. Maler, and A. Pnueli. On the analysis of dynamical systems having piecewise-constant derivatives. *Theoretical Computer Science*, 138:35–65, 1995.

[And68]   J. A. Anderson. A memory storage model utilizing spatial correlation functions. *Kybernetik*, 5:113–119, 1968.

[And72]   J. A. Anderson. A simple neural network generating an interactive memory. *Mathematical Biosciences*, 14:197–220, 1972.

[Art96]   Z. Artstein. Examples of stabilization with hybrid feedback. In
            R. Alur, T. A. Henzinger, and E. D. Sontag, editors, *Lecture Notes
            in Computer Science: Hybrid Systems III: Verification and Con-
            trol*, volume 1066, pages 173–181. Springer, New York, 1996.

[AS93a]   F. Albertini and E. D. Sontag. For neural networks, function
            determines form. *Neural Networks*, 6:975–990, 1993.

[AS93b]   F. Albertini and E. D. Sontag. Identifiability of discrete-time neu-
            ral networks. In *Proc. European Control Conference*, pages 460–
            465, Groningen, June 1993.

[AS94]   F. Albertini and E. D. Sontag. State observability in recurrent
            neural networks. *Systems and control Letters*, 22:235–244, 1994.

[Atk89]   K. E. Atkinson. *An Introduction to Numerical Analysis*. Wiley,
            New York, 1989.

[Bar92]   A. R. Barron. Neural net approximation. In *Proc. Seventh Yale
            Workshop on Adaptive and Learning Systems*, pages 69–72, New
            Haven, CT, 1992. Yale University Press.

[BB92]   A. Bertiaume and G. Brassard. The quantum challenge to struc-
            tural complexity theory. In *Proc. Seventh IEEE Conference on
            Structure in Complexity Theory*, pages 132–137, Boston, MA, June
            1992. IEEE Computer Society.

[BC96]   O. Bournez and M. Cosnard. On the computational power of dy-
            namical systems and hybrid systems. *Theoretical Computer Sci-
            ence*, 168(2):417–459, November 1996.

[BCSS98]   L. Blum, F. Cucker, M. Shub, and S. Smale. *Complexity and Real
            Computation*. Springer-Verlag, 1998.

[BDG90]   J. L. Balcázar, J. Díaz, and J. Gabarró. *Structural Complexity:
            volumes I and II, EATCS Monograph Series*. Springer-Verlag,
            Berlin, 1988-1990. (Second Edition for Volume I in 1995).

[Ben82a]   P. Benioff. Quantum mechanical Hamiltonian models of Turing
            machines. *Journal of Statistical Physics*, 29:515–546, 1982.

[Ben82b]   P. Benioff. Quantum mechanical Hamiltonian models of Tur-
            ing machines that dissipate no energy. *Physical Review Letters*,
            48:1581–1585, 1982.

[BG90]   G. L. Baker and J. P. Gollub. *Chaotic Dynamics: an Introduction*.
            Cambridge University Press, Cambridge, England, 1990.

[BGM93]   A. Back, J. Guckenheimer, and M. Myers. A dynamical simulation facility for hybrid systems. In R. L. Grossman, A. Nerode, A. P. Ravn, and H. Rischel, editors, *Lecture notes in Computer Science: Hybrid Systems*, volume 736, pages 255–267. Springer-Verlag, New York, 1993.

[BGS97]   J. L. Balcázar, R. Gavaldà, and H. T. Siegelmann. Computational power of neural networks: A characterization in terms of kolmogorov complexity. *IEEE Transactions on Information Theory*, 43(4):1175–1183, 1997.

[BGSS93]   J. L. Balcázar, R. Gavaldà, H. T. Siegelmann, and E. D. Sontag. Some structural complexity aspects of neural computation. In *Proc. Eighth IEEE Structure in Complexity Theory Conference*, pages 253–265, San Diego, CA, May 1993. IEEE Computer Society.

[BH89]   E. B. Baum and D. Haussler. What size net gives valid generalization? *Neural Computation*, 1:151–160, 1989.

[BHM92]   J. L. Balcázar, M. Hermo, and E. Mayordomo. Characterizations of logarithmic advice complexity classes. *Information Processing 92, IFIP Transactions A-12*, 1:315–321, 1992.

[Bra93]   M. S. Branicky. Topology of hybrid systems. In *Proc. Thirty-second IEEE Conference on Decision and Control*, pages 2309–2314, San Antonio, TX, 1993.

[Bra94]   M. S. Branicky. Analog computation with continuous ODEs. In *Proc. IEEE Workshop on Physics and Computation*, pages 265–274, Dallas, TX, 1994.

[Bra95]   M. S. Branicky. Universal computation and other capabilities of hybrid and continuous dynamical systems. *Theoretical Computer Science*, 138(1), 1995.

[Bro89]   R. W. Brockett. Smooth dynamical systems which realize arithmetical and logical operations. In H. Nijmeijer and L. M. Schumacher, editors, *Three Decades of Mathematical Systems Theory*, pages 19–30. Springer-Verlag, Berlin, 1989.

[Bro91]   R. W. Brockett. Dynamical systems that sort lists, diagonalize matrices, and solve linear programming problems. *Linear Algebra and its Applications*, 146:79–91, February 1991.

[Bro93]   R. W. Brockett. Hybrid models for motion control systems. In H. L. Trentelman and J. C. Willems, editors, *Essays in Control: Perspectives in the Theory and its Applications*, pages 29–53. Birkhäuser, Boston, 1993.

[BSS89]   L. Blum, M. Shub, and S. Smale. On a theory of computation and complexity over the real numbers: NP completeness, recursive functions, and universal machines. *Bulletin of the American Mathematical Society*, 21:1–46, 1989.

[Bus70]   V. Bush. *Pieces of the Action*. William Morrow, New York, 1970.

[BV93]   E. Bernstein and U. Vazirani. Quantum complexity theory. In *Proc. Twenty-fifth Annual ACM Symposium on the Theory of Computing*, pages 11–20, San Diego, CA, May 1993.

[Cas96]   M. P. Casey. The dynamics of discrete-time computation with application to recurrent neural networks and finite state machine extraction. *Neural Computation*, 8(6):1135–1178, 1996.

[CG97]   F. Cucker and D. Grigoriev. On the power of real Turing machines over binary inputs. *SIAM Journal on Computing*, 26:243–254, 1997.

[Cha74a]   G. Chaitin. Information theoretic limitations of formal systems. *Journal of the Association of Computer Machinery*, 21:403–424, 1974.

[Cha74b]   G. Chaitin. A theory of program size formally identical to information theory. Technical Report RC 4805, IBM, Yorktown Heights, NY, 1974.

[CSSM89]   A. Cleeremans, D. Servan-Schreiber, and J. McClelland. Finite state automata and simple recurrent recurrent networks. *Neural Computation*, 1(3):372, 1989.

[CSV84]   A. K. Chandra, L. Stockmeyer, and U. Vishkin. Constant depth reducibility. *SIAM Journal on Computing*, 13:423–439, 1984.

[Cyb89]   G. Cybenko. Approximation by superpositions of a sigmoidal function. *Mathematics of Control, Signals, and Systems*, 2:303–314, 1989.

[Deu85]   D. Deutsch. Quantum theory, the Church-Turing principle and the universal quantum computer. *Proceedings of the Royal Society of London*, A400:96–117, 1985.

[Deu89]   D. Deutsch. Quantum computational networks. *Proceedings of the Royal Society of London*, A425:73–90, 1989.

[DO77a]   R. L. Dobrushin and S. I. Ortyukov. Lower bound for the redundancy of self-correcting arrangement of unreliable functional elements. *Problems of Information Transmission*, 13:59–65, 1977.

[DO77b]   R. L. Dobrushin and S. I. Ortyukov. Upper bound for the redundancy of self-correcting arrangement of unreliable functional elements. *Problems of Information Transmission*, 13:346–353, 1977.

[Elm90]   J. L. Elman. Finding structure in time. *Cognitive Science*, 14:179–211, 1990.

[Fay91]   L. Faybusovich. Hamiltonian structure of dynamical systems which solve linear programming problems. *Physica D*, 53:217–232, 1991.

[Fey82]   R. P. Feynman. Simulating physics with computers. *International Journal of Theoretical Physics*, 21(6/7):467–488, 1982.

[Fey86]   R. P. Feynman. Quantum mechanical computers. *Foundations of Physics*, 16(6):507–531, 1986. Originally appeared in Optics News, February 1985.

[FG90]   S. Franklin and M. Garzon. Neural computability. In O. M. Omidvar, editor, *Progress in Neural Networks*, pages 128–144. Ablex, Norwood, NJ, 1990.

[Fin94]   Y. Finkelstein. *Cholinergic Mechanisms of Control and Adaptation in the Rat Septo-Hippocampus under Stress Conditions*. PhD thesis, Hebrew University in Jerusalem, Israel, 1994.

[Fra89]   J. A. Franklin. On the approximate realization of continuous mappings by neural networks. *Neural Networks*, 2:183–192, 1989.

[FSS81]   M. Furst, J. B. Saxe, and M. Sipser. Parity, circuits, and the polynomial-time hierarchy. In *Proc. Twenty-second IEEE Symposium on Foundations of Computer Science*, pages 260–270, 1981.

[GF89]   M. Garzon and S. Franklin. Neural computability. In *Proc. Third International Joint Conference on Neural Networks*, volume 2, pages 631–637, 1989.

[GH83]   J. Guckenheimer and P. Holmes. *Nonlinear Oscillations, Dynamical Systems, and Bifurcations of Vector Fields*. Springer-Verlag, New York, 1983.

[GJ79]   M. R. Garey and D. S. Johnson. *Computers and Intractability: A Guide to the Theory of NP-Completeness.* Freeman, New York, 1979.

[GMC⁺92] C. L. Giles, C. B. Miller, D. Chen, H. H. Chen, G. Z. Sun, and Y. C. Lee. Learning and extracting finite state automata with second-order recurrent neural networks. *Neural Computation*, 4(3):393–405, 1992.

[GNRe93] R. L. Grossman, A. Nerode, A. P. Ravn, and H. Rischel editors. *Lecture Notes in Computer Science: Hybrid Systems*, volume 736. Springer-Verlag, New York, 1993.

[GOY87]   C. Grebogi, E. Ott, and J. A. Yorke. Chaos, strange attractors, and fractal basin boundaries in nonlinear dynamics. *Science*, 238:632–637, October 1987.

[GS98]   R. Gavaldà and H. T. Siegelmann. Discontinuities in recurrent neural networks. *Neural Computation*, 1998. To appear.

[Her96]   M. Hermo. *Nonuniform complexity classes with sub-linear advice functions.* PhD thesis, Universidad del Pais Vasco, Donostia, Spain, 1996.

[HKP91]   J. Hertz, A. Krogh, and R. Palmer. *Introduction to the Theory of Neural Computation.* Addison-Wesley, Redwood City, 1991.

[HM94]   U. Helmke and J. B. Moore. *Optimization and Dynamical Systems.* Springer-Verlag, London, 1994.

[Hon88]   J. W. Hong. On connectionist models. *Comm. Pure and Applied Mathematics*, 41:1039–1050, 1988.

[Hop84]   J. J. Hopfield. Neurons with graded responses have collective computational properties like those of two-state neurons. *Proceedings of the National Academy of Sciences*, 81:3088–3092, 1984.

[Hor91]   K. Hornik. Approximation capabilities of multilayer feedforward networks. *Neural Networks*, 4:251–257, 1991.

[HS74]   M. Hirsch and S. Smale. *Differential Equations, Dynamical Systems and Linear Algebra.* Academic Press, New York, 1974.

[HS87]   R. Hartley and H. Szu. A comparison of the computational power of neural network models. In *Proc. IEEE Conference on Neural Networks*, pages 17–22, 1987.

[HSW90]   K. Hornik, M. Stinchcombe, and H. White. Universal approxima-
          tion of an unknown mapping and its derivatives using multilayer
          feedforward networks. *Neural Networks*, 3:551–560, 1990.

[HT85]   J. J. Hopfield and D. W. Tank. Neural computation of decisions in
          optimization problems. *Biological Cybernetics*, 52:141–152, 1985.

[HU79]   J. E. Hopcroft and J. D. Ullman. *Introduction to Automata The-
          ory, Languages, and Computation.* Addison-Wesley, Reading, MA,
          1979.

[Ind95]   P. Indyk. Optimal simulation of automata by neural nets. In
          Ernst W. Mayr and Claude Puech, editors, *Lecture Notes in
          Computer Science: Proc. Twelfth Annual Symposyum on Theo-
          retical Aspects of Computer Science*, volume 900, pages 337–348.
          Springer, Munich, March 1995.

[Kar84]   N. Karmarkar. A new polynomial time algorithm in linear pro-
          gramming. *Combinatorica*, 4:373–395, 1984.

[Kar91]   H. Karloff. *Linear Programming.* Birkhäuser, Boston, MA, 1991.

[KCG94]   P. Koiran, M. Cosnard, and M. Garzon. Computability with low-
          dimensional dynamical systems. *Theoretical Computer Science*,
          132(1-2):113–128, September 1994.

[KHD94]   J. P. Keating, J. H. Hannay, and A. M. O. Dealmeida. Optical
          realization of the Baker's transformation. *Nonlinearity*, 7(5):1327–
          1342, 1994.

[KL80]   R. M. Karp and R. J. Lipton. Some connections between uni-
          form and nonuniform complexity classes. In *Proc. Twelfth ACM
          Symposium on Theory of Computing*, pages 302–309, 1980.

[KL82]   R. M. Karp and R. Lipton. Turing machines that take advice.
          Enseignment Mathematique, 28:191–209, 1982.

[Kle56]   S. C. Kleene. Representation of events in nerve nets and finite
          automata. In C. E. Shannon and J. McCarthy, editors, *Automata
          Studies*, pages 3–42. Princeton University Press, Princeton, NJ,
          1956.

[Ko87]   K. Ko. On helping by robust oracle machines. *Theoretical Com-
          puter Science*, 52:15–36, 1987.

[Kob81]   K. Kobayashi. On compressibility of infinite sequences. Technical
          Report C-34, Department of Information Sciences, Tokyo Institute
          of Technology, 1981.

[Koh72]   T. Kohonen. Correlation matrix memories. *IEEE Transactions on Computers*, 21:353–359, 1972.

[Koi93]   P. Koiran. A weak version of the Blum, Shub & Smale model. In *Proc. Thirty-fourth IEEE Symposium on Foundations of Computer Science*, pages 486–495, 1993.

[Kol65]   A. N. Kolmogorov. Three approaches to the quantitative definition of information. *Problems of Information Transmission*, 1(1):1–7, 1965.

[KS96]   J. Kilian and H. T. Siegelmann. The dynamic universality of sigmoidal neural networks. *Information and Computation*, 128(1):48–56, July 1996.

[Lip95]   R. J. Lipton. DNA solution of hard computational problems. *Science*, 268(5210):542–545, April 28 1995.

[Lit74]   W. A. Little. The existence of persistent states in the brain. *Mathematical Biosciences*, 19:101–120, 1974.

[Lov69]   D. W. Loveland. A variant of the Kolmogorov concept of complexity. *Information and Control*, 15:115–133, 1969.

[LS78]   W. A. Little and G. L. Shaw. Analytic study of the memory storage capacity of a neural network. *Mathematical Biosciences*, 39:281–290, 1978.

[LV90]   M. Li and P. M. B. Vitányi. Kolmogorov complexity and its applications. In *Handbook of Theoretical Computer Science*, volume A, pages 187–254. Elsevier, Amsterdam, 1990. Reprinted MIT Press, Cambridge, MA, 1994.

[LV93]   M. Li and P. M. B. Vitányi. *An Introduction to Kolmogorov Complexity and its Applications: Texts and Monographs in Computer Science Series*. Springer-Verlag, New York, 1993.

[LZ76]   A. Lempel and J. Ziv. On the complexity of finite sequences. *IEEE Transactions on Information Theory*, 22(1):75–81, 1976.

[Mar66]   P. Martin-Löf. The definition of random sequences. *Information and Control*, 9(6):602–619, December 1966.

[Mar71]   P. Martin-Löf. Complexity oscillations in infinite binary sequences. *Wahrscheinlichkeit verw. Geb.*, 19:225–230, 1971.

[MG62] A. A. Muchnik and S. G. Gindikin. The completeness of a system made up of non-reliable elements realizing a function of algebraic logic. *Soviet Physics Doklady*, 7:477–479, 1962.

[Min67] M. L. Minsky. *Computation: Finite and Infinite Machines.* Prentice Hall, Engelwood Cliffs, NJ, 1967.

[MM97] Klaus Meer and Christian Michaux. A survey on real structural complexity theory. *Bulletin of the Belgium Mathematical Society*, 4:113–148, 1997.

[Moo90] C. Moore. Unpredictability and undecidability in dynamical systems. *Physical Review Letters*, 64:2354–2357, 1990.

[Moo91] C. Moore. Generalized shifts: unpredictability and undecidability in dynamical systems. *Nonlinearity*, 4:199–230, 1991.

[Mor93] M. Morita. Associative memory with nonmonotonic dynamics. *Neural Networks*, 6:115–126, 1993.

[MP43] W. S. McCulloch and W. Pitts. A logical calculus of the ideas immanent in nervous activity. *Bulletin of Mathematical Biophysics*, 5:115–133, 1943.

[MS98] W. Maass and E. D. Sontag. Analog neural nets with gaussian or other common noise distribution cannot recognize arbitrary regular languages. *Neural Computation*, 1998. submitted.

[MSS91] W. Maass, G. Schnitger, and E. D. Sontag. On the computational power of sigmoid versus Boolean threshold circuits. In *Proc. Thirty-second IEEE Symposium on Foundations of Computer Science*, pages 767–776, 1991.

[Mul56] D. E. Muller. Complexity in electronic switching circuits. *IRE Transactions on Electronic Computers*, EC-5:15–19, 1956.

[Mur71] S. Muroga. *Threshold Logic and its Applications.* Wiley, New York, 1971.

[Nak72] K. Nakano. Associatron — a model of associative memory. *IEEE Transactions on Systems, Man and Cybernetics*, 2:380–388, 1972.

[NK91] J. Nyce and P. Kahn. *From Memex to Hypertext: Vannevar Bush and Mind's Machine.* Academic Press, San Diego, CA, 1991.

[NK93] A. Nerode and W. Kohn. Models for hybrid systems: Automata, topologies, controllability, observability. In A. P. Ravn

R. L. Grossman, A. Nerode and H. Rischel, editors, *Lecture Notes in Computer Science: Hybrid Systems*, volume 736, pages 317–356. Springer-Verlag, New York, 1993.

[NN71]   K. Nakano and J-I Nagumo. Information processing using a model of associative memory. In *Proc. Second International Conference on Artificial Intelligence*, pages 101–106, High Wycomb, England, 1971. University Microfilms.

[NSCA97]   J. P. Neto, H. T. Siegelmann, J. F. Costa, and C. P. Araujo. Turing universality of neural nets (revisited). In F. Pichler and R. Moreno Diaz, editors, *Lecture notes in Computer Science: Computer-Aided Systems Theory*. Springer-Verlag, 1997.

[NY94]   H. E. Nusse and J. A. Yorke. *Dynamics: Numerical Explorations*. Springer-Verlag, New York, 1994.

[Nyc92]   J. Nyce. Analogy or identity: Brain and machine at the Macy conferences on cybernetics. *SIGBIO Newsletter: Published by the Association for Computing Machinery, Special Interest Group on Biomedial Computing*, 12:32–37, 1992.

[OG96]   C. W. Omlin and C. L. Giles. Constructing deterministic finite-state automata in recurrent neural networks. *Journal of the Association of Computing Machinery*, 45(6):937–972, 1996.

[OM96]   P. Orponen and M. Matamala. Universal computation by finite two-dimensional coupled map lattices. In *Proc. Physics and Computation*, pages 243–247, Boston, MA, November 1996.

[OM98]   P. Orponen and W. Maass. On the effect of analog noise on discrete time analog computations. *Neural Computation*, 1998. to appear.

[Omo84]   S. Omohundro. Modeling cellular automata with partial differential equations. *Physica D*, 10:128–134, 1984.

[Orp94]   P. Orponen. Computational complexity of neural networks: A survey. *Nordic Journal of Computing*, 1:94–110, 1994. Previous version in Proc. Seventeenth Symposium on Mathematical Foundations of Computer Science, 50-61, 1992.

[Ort78]   S. I. Ortyukov. Synthesis of asymptotically nonredundant self-correcting arrangements of unreliable functional elements. *Problems of Information Transmission*, 13:247–251, 1978.

[Ott93]   E. Ott. *Chaos in Dynamical Systems*. Cambridge University Press, Cambridge, England, 1993.

[Par94]   I. Parberry. *Circuit Complexity and Neural Networks*. MIT Press, Cambridge, MA, 1994.

[Paz71]   A. Paz. *Introduction to Probabilistic Automata*. Academic Press, New York, 1971.

[Pen89]   R. Penrose. *The Emperor's New Mind*. Oxford University Press, Oxford, England, 1989.

[Pen94]   R. Penrose. *Shadows of the Mind*. Oxford University Press, Oxford, England, 1994.

[Pip88]   N. Pippenger. Reliable computation by formulae in the presence of noise. *IEEE Transactions on Information Theory*, 34:194–197, 1988.

[Pip89]   N. Pippenger. Invariance of complexity measure of networks with unreliable gates. *Journal of the Association of Computing Machinery*, 36:531–539, 1989.

[Pip90]   N. Pippenger. Developments in: The synthesis of reliable organisms from unreliable components. In *Proc. Symposia in Pure Mathematics*, volume 5, pages 311–324, 1990.

[Pol87]   J. B. Pollack. *On Connectionist Models of Natural Language Processing*. PhD thesis, Computer Science Dept., Univ. of Illinois, Urbana, 1987.

[Pou74]   M. B. Pour-El. Abstract computability and its relation to the general purpose analog computer (some connections between logic, differential equations and analog computers). *Transactions of the American Mathematical Society*, 199:1–29, 1974.

[PR88]   M. B. Pour-El and J. I. Richards. *Computability in Analysis and Physics*. Springer-Verlag, New York, 1988.

[PS82]   C. H. Papadimitriou and K. Steiglitz. *Combinatorial Optimization*. Prentice Hall, Englewood Cliffs, NJ, 1982.

[Rab63]   M. Rabin. Probabilistic automata. *Information and Control*, 6:230–245, 1963.

[Rub81]   L. A. Rubel. A universal differential equation. *Bulletin of the American Mathematical Society*, 4(3):345–349, May 1981.

[Rub89]   L. A. Rubel. Digital simulation of analog computation and Church's thesis. *The Journal of Symbolic Logic*, 54(3):1011–1017, September 1989.

[Rub93]   L. A. Rubel. The extended analog computer. *Advances in Applied Mathematics*, 14:39–50, 1993.

[Sav76]   J. E. Savage. *The Complexity of Computing*. Wiley, New York, 1976.

[SBF98]   H. T. Siegelmann, A. Ben-Hur, and S. Fishman. A theory of complexity for continuous time dynamics. 1998. Submitted.

[Sch73]   C. P. Schnorr. The process complexity and effective random tests. *Journal of Computer and System Sciences*, 7(4):376–388, August 1973.

[Sch86]   U. Schöning. *Lecture notes in Computer Science: Complexity and Structure*, volume 211. Springer-Verlag, New York, 1986.

[SCLG91]   G. Z. Sun, H. H. Chen, Y. C. Lee, and C. L. Giles. Turing equivalence of neural networks with second order connection weights. In *Proc. IEEE International Joint Conference on Neural Networks*, volume 2, pages 357–362, 1991.

[SF98]   H. T. Siegelmann and S. Fishman. Analog computation with dynamical systems. *Physica D*, 1998. to appear.

[SG97]   H. T. Siegelmann and C. L. Giles. The complexity of language recognition by neural networks. *Neurocomputing*, 15:327–345, 1997.

[Sha41]   C. E. Shannon. Mathematical theory of the differential analyzer. *Journal of Mathematics and Physics of the Massachusetts Institute of Technology*, 20:337–354, 1941.

[Sha48]   C. E. Shannon. A mathematical theory of communication. *AT&T Bell Laboratories Technical Journal*, 27:379–423, 623–656, 1948.

[Sha56]   C. E. Shannon. A universal Turing machine with two internal states. In C. E. Shannon and J. McCarthy, editors, *Automata Studies*, pages 156–165. Princeton University Press, Princeton, NJ, 1956.

[Sho94]   P. W. Shor. Algorithms for quantum computation: Discrete logarithms and factoring. In S. Goldwasser, editor, *Proc. Thirty-fifth Annual Symposium on Foundations of Computer Science*, pages 124–134, Los Alamitos, CA, November 1994. IEEE Computer Society Press.

[SI71]   A. R. Smith III. Simple computation-universal cellular spaces. *Journal of the Association of Computing Machinery*, 18:339, 1971.

[Sie95]   H. T. Siegelmann. Computation beyond the Turing limit. *Science*, 268(5210):545–548, April 28 1995.

[Sie96a]   H. T. Siegelmann. Neural networks and finite automata. *Journal of Computational Intelligence*, 12(4), 1996.

[Sie96b]   H. T. Siegelmann. On NIL: The software constructor of neural networks. *Parallel Processing Letters*, 6(4):575–582, 1996.

[Sie98]   H. T. Siegelmann. Neural dynamics with stochasticity. In *Adaptive Processing of Sequences and Data Structures*, pages 346–369. Springer, 1998.

[SM98]   H. T. Siegelmann and M. Margernstern. Nine switch-affine neurons suffice for Turing universality. *Neural Networks*, 1998. submitted.

[Sma91]   S. Smale. Theory of computation. In *Proc. Symp. on the Current State and Prospects of Mathematics*, pages 59–69, Barcelona, June 1991.

[Son90]   E. D. Sontag. *Mathematical Control Theory: Deterministic Finite Dimensional Systems*. Springer-Verlag, New York, 1990.

[Son93]   E. D. Sontag. Neural networks for control. In H. L. Trentelman and J. C. Willems, editors, *Essays on Control: Perspectives in the Theory and its Applications*. Birkhäuser, Cambridge, MA, 1993.

[SR98]   H. T. Siegelmann and A. Roitershtein. Noisy analog neural networks and definite languages: stochastic kernels approach. Technical report, Technion, Haifa, Israel, 1998.

[SS91]   H. T. Siegelmann and E. D. Sontag. Turing computability with neural nets. *Applied Mathematics Letters*, 4(6):77–80, 1991.

[SS94]   H. T. Siegelmann and E. D. Sontag. Analog computation via neural networks. *Theoretical Computer Science*, 131:331–360, 1994.

[SS95]   H. T. Siegelmann and E. D. Sontag. On computational power of neural networks. *Journal of Computer and System Sciences*, 50(1):132–150, 1995.

[Sus92]   H. J. Sussmann. Uniqueness of the weights for minimal feedforward nets with a given input-output map. *Neural Networks*, 5:589–593, 1992.

[SW90]   M. Stinchcombe and H. White. Approximating and learning
           unknown mappings using multilayer feedforward networks with
           bounded weights. In *Proc. IEEE International Joint Conference
           on Neural Networks*, volume 3, pages 7–16, 1990.

[Tav87]   L. Tavernini. Differential automata and their discrete simula-
           tors. *Nonlinear Analysis, Theory, Methods and Applications*,
           11(6):665–683, 1987.

[TM87]   T. Toffoli and N. Margolus. *Cellular Automata Machines*. MIT
           Press, Cambridge, MA, 1987.

[Tur36]   A. M. Turing. On computable numbers. *Proceedings of the London
           Mathematical Society (2)*, 42:230–265, 1936.

[Uli74]   D. Ulig. On the synthesis of self-correcting schemes from func-
           tional elements with a small number of reliable elements. *Math.
           Notes. Acad. Sci. USSR*, 15:558–562, 1974.

[vN51]   J. von Neumann. Various techniques use in connection with ran-
           dom digits (notes by g.e. forsythe, national bureau of standards).
           In *Applied Math Series*, volume 12, pages 36–38. 1951.

[vN56]   J. von Neumann. Probabilistic, logics and the synthesis of re-
           liable organisms from unreliable components. In C. E. Shannon
           and J. McCarthy, editors, *Automata Studies*. Princeton University
           Press, Princeton, NJ, 1956.

[WBL69]   D. J. Willshaw, O. P. Buneman, and H. C. Longuet-Higgins. Non-
           holographic associative memory. *Nature*, 222:960–962, 1969.

[Wie49]   N. Wiener. *Extrapolation, interpolation, and smoothing of station-
           ary time series*. MIT Press, Cambridge, MA, 1949.

[Wol91]   D. Wolpert. A computationally universal field computer which
           is purely linear. Technical Report LA-UR-91-2937, Los Alamos
           National Laboratory, Los Alamos, NM, 1991.

[WZ89]   R. J. Williams and D. Zipser. A learning algorithm for continu-
           ally running fully recurrent neural networks. *Neural Computation*,
           1(2):270–280, 1989.

[Zac82]   S. Zachos. Robustness of probabilistic computational complexity
           classes under definitional perturbations. *Information and Control*,
           54:143–154, 1982.

Index

Progress in Theoretical Computer Science

Editor

Ronald V. Book
Department of Mathematics
University of California
Santa Barbara, CA 93106

Editorial Board

Progress in Theoretical Computer Science is a series that focuses on the theoretical aspects of computer science and on the logical and mathematical foundations of computer science, as well as the applications of computer theory. It addresses itself to research workers and graduate students in computer and information science departments and research laboratories, as well as to departments of mathematics and electrical engineering where an interest in computer theory is found.

The series publishes research monographs, graduate texts, and polished lectures from seminars and lecture series. We encourage preparation of manuscripts in some form of TeX for delivery in camera-ready copy, which leads to rapid publication, or in electronic form for interfacing with laser printers or typesetters.

Proposals should be sent directly to the Editor, any member of the Editorial Board, or to: Birkhäuser Boston, 675 Massachusetts Ave., Cambridge, MA 02139. The Series includes:

1. Leo Bachmair, *Canonical Equational Proofs*
2. Howard Karloff, *Linear Programming*
3. Ker-I Ko, *Complexity Theory of Real Functions*
4. Guo-Qiang Zhang, *Logic of Domains*
5. Thomas Streicher, *Semantics of Type Theory: Correctness, Completeness and Independence Results*
6. Julian Charles Bradfield, *Verifying Temporal Properties of Systems*
7. Alistair Sinclair, *Algorithms for Random Generation and Counting*
8. Heinrich Hussmann, *Nondeterminism in Algebraic Specifications and Algebraic Programs*
9. Pierre-Louis Curien, *Categorical Combinators, Sequential Algorithms and Functional Programming*
10. J. Köbler, U. Schöning, and J. Torán, *The Graph Isomorphism Problem: Its Structural Complexity*
11. Howard Straubing, *Finite Automata, Formal Logic, and Circuit Complexity*
12. Dario Bini and Victor Pan, *Polynomial and Matrix Computations, Volume 1 Fundamental Algorithms*
13. James S. Royer and John Case, *Subrecursive Programming Systems: Complexity & Succinctness*